# PSAT Subject Test

## Mathematics

### Student Practice Workbook

### + Two Full-Length PSAT Math Tests

**Math Notion**

**www.MathNotion.com**

PSAT Subject Test Mathematics

Published in the United State of America By

The Math Notion

Web: WWW.MathNotion.com

Email: info@Mathnotion.com

**SCAN ME**

ISBN: 978-1-63620-047-7

# The Math Notion

**Michael Smith** has been a math instructor for over a decade now. He launched the Math Notion. Since 2006, we have devoted our time to both teaching and developing exceptional math learning materials. As a test prep company, we have worked with thousands of students. We have used the feedback of our students to develop a unique study program that can be used by students to drastically improve their math scores fast and effectively. We have more than a thousand Math learning books including:

- **SAT Math Prep**
- **ACT Math Prep**
- **GRE Math Prep**
- **Accuplacer Math Prep**
- **Common Core Math Prep**
- **many Math Education Workbooks, Study Guides, Practice and Exercise Books**

As an experienced Math test preparation company, we have helped many students raise their standardized test scores—and attend the colleges of their dreams: We tutor online and in person, we teach students in large groups, and we provide training materials and textbooks through our website and through Amazon.

You can contact us via email at:

info@Mathnotion.com

## Get the Targeted Practice You Need to Ace the PSAT Math Test!

**PSAT Subject Test - Mathematics** includes easy-to-follow instructions, helpful examples, and plenty of math practice problems to assist students to master each concept, brush up their problem-solving skills, and create confidence.

The PSAT math practice book provides numerous opportunities to evaluate basic skills along with abundant remediation and intervention activities. It is a skill that permits you to quickly master intricate information and produce better leads in less time.

Students can boost their test-taking skills by taking the book's two practice PSAT Math exams. All test questions answered and explained in detail.

**Important Features of the PSAT Math Book:**

- A **complete review** of PSAT math test topics,
- Over 2,500 practice problems covering all topics tested,
- The most important concepts you need to know,
- Clear and concise, easy-to-follow sections,
- Well designed for enhanced learning and interest,
- Hands-on experience with all question types
- **2 full-length practice tests** with detailed answer explanations
- Cost-Effective Pricing

Powerful math exercises to help you avoid traps and pacing yourself to beat the PSAT test. Students will gain valuable experience and raise their confidence by taking math practice tests, learning about test structure, and gaining a deeper understanding of what is tested on the PSAT Math. If ever there was a book to respond to the pressure to increase students' test scores, this is it.

# WWW.MathNotion.COM

## … So Much More Online!

✓ FREE Math Lessons

✓ More Math Learning Books!

✓ Mathematics Worksheets

✓ Online Math Tutors

## For a PDF Version of This Book

SCAN ME

Please Visit WWW.MathNotion.com

# Contents

## PSAT Subject Test – Mathematics

# Chapter 1 :

# Integers and Number Theory

## Topics that you'll practice in this chapter:

- ✓ Rounding
- ✓ Whole Number Addition and Subtraction
- ✓ Whole Number Multiplication and Division
- ✓ Rounding and Estimates
- ✓ Adding and Subtracting Integers
- ✓ Multiplying and Dividing Integers
- ✓ Order of Operations
- ✓ Ordering Integers and Numbers
- ✓ Integers and Absolute Value
- ✓ Factoring Numbers
- ✓ Greatest Common Factor (GCF)
- ✓ Least Common Multiple (LCM)

*"Whereuer there is number, there is beauty."* –*Proclus*

# Rounding

✍ **Round each number to the nearest ten.**

1) 42 = ____          5) 19 = ____          9) 48 = ____

2) 88 = ____          6) 25 = ____          10) 81 = ____

3) 24 = ____          7) 93 = ____          11) 58 = ____

4) 57 = ____          8) 71 = ____          12) 87 = ____

✍ **Round each number to the nearest hundred.**

13) 198 = ____       17) 321 = ____       21) 580 = ____

14) 387 = ____       18) 433 = ____       22) 868 = ____

15) 816 = ____       19) 579 = ____       23) 480 = ____

16) 101 = ____       20) 825 = ____       24) 287 = ____

✍ **Round each number to the nearest thousand.**

25) 1,382 = ____     29) 9,099 = ____     33) 52,866 = ____

26) 3,420 = ____     30) 22,980 = ____    34) 85,190 = ____

27) 4,254 = ____     31) 45,188 = ____    35) 70,990 = ____

28) 6,861 = ____     32) 16,808 = ____    36) 26,869 = ____

# Rounding and Estimates

✎ **Estimate the sum by rounding each number to the nearest ten.**

1) $13 + 22 =$ _____

2) $71 + 23 =$ _____

3) $61 + 58 =$ _____

4) $56 + 85 =$ _____

5) $368 + 249 =$ _____

6) $330 + 903 =$ _____

7) $471 + 293 =$ _____

8) $1,950 + 2,655 =$ _____

✎ **Estimate the product by rounding each number to the nearest ten.**

9) $32 \times 71 =$ _____

10) $12 \times 33 =$ _____

11) $31 \times 83 =$ _____

12) $19 \times 11 =$ _____

13) $42 \times 76 =$ _____

14) $63 \times 34 =$ _____

15) $19 \times 31 =$ _____

16) $59 \times 71 =$ _____

✎ **Estimate the sum or product by rounding each number to the nearest ten.**

$$
17)\ \begin{array}{r} 29 \\ \times\ 12 \\ \hline \phantom{00} \end{array}
\qquad
19)\ \begin{array}{r} 48 \\ +\ 82 \\ \hline \phantom{00} \end{array}
\qquad
21)\ \begin{array}{r} 37 \\ \times\ 14 \\ \hline \phantom{00} \end{array}
$$

$$
18)\ \begin{array}{r} 37 \\ \times\ 26 \\ \hline \phantom{00} \end{array}
\qquad
20)\ \begin{array}{r} 65 \\ +44 \\ \hline \phantom{00} \end{array}
\qquad
22)\ \begin{array}{r} 71 \\ +\ 32 \\ \hline \phantom{00} \end{array}
$$

# Adding and Subtracting Integers

✎ **Find each sum.**

1) $14 + (-6) =$

2) $(-13) + (-20) =$

3) $5 + (-28) =$

4) $50 + (-12) =$

5) $(-7) + (-15) + 3 =$

6) $30 + (-14) + 8 =$

7) $40 + (-10) + (-14) + 17 =$

8) $(-15) + (-20) + 13 + 35 =$

9) $40 + (-20) + (38 - 29) =$

10) $28 + (-12) + (30 - 12) =$

✎ **Find each difference.**

11) $(-18) - (-7) =$

12) $25 - (-14) =$

13) $(-20) - 36 =$

14) $34 - (-19) =$

15) $51 - (30 - 21) =$

16) $17 - (5) - (-24) =$

17) $(35 + 20) - (-46) =$

18) $48 - 16 - (-8) =$

19) $62 - (28 + 17) - (-15) =$

20) $58 - (-23) - (-31) =$

21) $19 - (-8) - (-13) =$

22) $(19 - 24) - (-14) =$

23) $27 - 33 - (-21) =$

24) $58 - (32 + 24) - (-9) =$

25) $36 - (-30) + (-17) =$

26) $27 - (-42) + (-31) =$

# Multiplying and Dividing Integers

✍ **Find each product.**

1) $(-9) \times (-5) =$

2) $(-3) \times 9 =$

3) $8 \times (-12) =$

4) $(-7) \times (-20) =$

5) $(-3) \times (-5) \times 6 =$

6) $(14 - 3) \times (-8) =$

7) $12 \times (-9) \times (-3) =$

8) $(140 + 10) \times (-2) =$

9) $10 \times (-12 + 8) \times 3 =$

10) $(-8) \times (-5) \times (-10) =$

✍ **Find each quotient.**

11) $42 \div (-7) =$

12) $(-48) \div (-6) =$

13) $(-40) \div (-8) =$

14) $54 \div (-2) =$

15) $152 \div 19 =$

16) $(-144) \div (-12) =$

17) $180 \div (-10) =$

18) $(-312) \div (-12) =$

19) $221 \div (-13) =$

20) $(-126) \div (6) =$

21) $(-161) \div (-7) =$

22) $-266 \div (-14) =$

23) $(-120) \div (-4) =$

24) $270 \div (-18) =$

25) $(-208) \div (-8) =$

26) $(135) \div (-15) =$

# Order of Operations

✎ **Evaluate each expression.**

1) $7 + (5 \times 4) =$

2) $14 - (3 \times 6) =$

3) $(19 \times 4) + 16 =$

4) $(16 - 7) - (8 \times 2) =$

5) $27 + (18 \div 3) =$

6) $(18 \times 8) \div 6 =$

7) $(32 \div 4) \times (-2) =$

8) $(9 \times 4) + (32 - 18) =$

9) $24 + (4 \times 3) + 7 =$

10) $(36 \times 3) \div (2 + 2) =$

11) $(-7) + (12 \times 3) + 11 =$

12) $(8 \times 5) - (24 \div 6) =$

13) $(7 \times 6 \div 3) - (12 + 9) =$

14) $(13 + 5 - 14) \times 3 - 2 =$

15) $(20 - 14 + 30) \times (64 \div 4) =$

16) $32 + \big(28 - (36 \div 9)\big) =$

17) $(7 + 6 - 4 - 7) + (15 \div 5) =$

18) $(85 - 20) + (20 - 18 + 7) =$

19) $(20 \times 2) + (14 \times 3) - 22 =$

20) $18 + 5 - (30 \times 3) + 20 =$

# Ordering Integers and Numbers

✎ **Order each set of integers from least to greatest.**

1) $8, -10, -5, -3, 4$        ___, ___, ___, ___, ___, ___

2) $-10, -18, 6, 14, 27$        ___, ___, ___, ___, ___, ___

3) $15, -8, -21, 21, -23$        ___, ___, ___, ___, ___, ___

4) $-14, -40, 23, -12, 47$        ___, ___, ___, ___, ___, ___

5) $59, -54, 32, -57, 36$        ___, ___, ___, ___, ___, ___

6) $68, 26, -19, 47, -34$        ___, ___, ___, ___, ___, ___

✎ **Order each set of integers from greatest to least.**

7) $18, 36, -16, -18, -10$        ___, ___, ___, ___, ___, ___

8) $27, 34, -12, -24, 94$        ___, ___, ___, ___, ___, ___

9) $50, -21, -13, 42, -2$        ___, ___, ___, ___, ___, ___

10) $37, 46, -20, -16, 86$        ___, ___, ___, ___, ___, ___

11) $-18, 88, -26, -59, 75$        ___, ___, ___, ___, ___, ___

12) $-65, -30, -25, 3, 14$        ___, ___, ___, ___, ___, ___

## Integers and Absolute Value

✍ **Write absolute value of each number.**

1) $|-2| =$

2) $|-27| =$

3) $|-20| =$

4) $|14| =$

5) $|6| =$

6) $|-55| =$

7) $|16| =$

8) $|2| =$

9) $|54| =$

10) $|-4| =$

11) $|-11|$

12) $|88| =$

13) $|0| =$

14) $|79| =$

15) $|-32| =$

16) $|-17| =$

17) $|42| =$

18) $|-46| =$

19) $|1| =$

20) $|-40| =$

✍ **Evaluate the value.**

21) $|-5| - \frac{|-21|}{7} =$

22) $14 - |3 - 15| - |-4| =$

23) $\frac{|-32|}{4} \times |-4| =$

24) $\frac{|7 \times (-3)|}{7} \times \frac{|-19|}{3} =$

25) $|4 \times (-5)| + \frac{|-40|}{5} =$

26) $\frac{|-45|}{9} \times \frac{|-24|}{12} =$

27) $|-12 + 8| \times \frac{|-7 \times 7|}{7} =$

28) $\frac{|-11 \times 2|}{4} \times |-16| =$

# Factoring Numbers

✎ **List all positive factors of each number.**

1) 9

2) 16

3) 24

4) 30

5) 26

6) 46

7) 20

8) 68

9) 28

10) 98

11) 14

12) 54

13) 55

14) 18

15) 63

16) 34

17) 50

18) 62

19) 95

20) 64

21) 70

22) 45

23) 22

24) 65

## Greatest Common Factor

✎ Find the GCF for each number pair.

| | | |
|---|---|---|
| 1) 6, 2 | 9) 12, 18 | 17) 42, 14 |
| 2) 4, 5 | 10) 4, 36 | 18) 16, 40 |
| 3) 3, 12 | 11) 6, 10 | 19) 9, 2, 3 |
| 4) 7, 3 | 12) 28, 52 | 20) 5, 15, 10 |
| 5) 5, 10 | 13) 25, 10 | 21) 7, 9, 2 |
| 6) 8, 48 | 14) 22, 24 | 22) 16, 64 |
| 7) 6, 18 | 15) 9, 54 | 23) 30, 48 |
| 8) 9, 15 | 16) 8, 54 | 24) 36, 63 |

# Least Common Multiple

✎ **Find the LCM for each number pair.**

1) 6, 9

2) 15, 45

3) 16, 40

4) 12, 36

5) 18, 27

6) 14, 42

7) 6, 30

8) 8, 56

9) 7, 21

10) 8, 20

11) 15, 25

12) 7, 9

13) 4, 11

14) 8, 28

15) 28, 56

16) 40, 50

17) 12, 13

18) 22, 11

19) 36, 20

20) 15, 35

21) 18, 81

22) 30, 54

23) 18, 45

24) 75, 25

# Answers of Worksheets

**Rounding**

| | | | |
|---|---|---|---|
| 1) 40 | 10) 80 | 19) 600 | 28) 7,000 |
| 2) 90 | 11) 60 | 20) 800 | 29) 9,000 |
| 3) 20 | 12) 90 | 21) 600 | 30) 23,000 |
| 4) 60 | 13) 200 | 22) 900 | 31) 45,000 |
| 5) 20 | 14) 400 | 23) 500 | 32) 17,000 |
| 6) 30 | 15) 800 | 24) 300 | 33) 53,000 |
| 7) 90 | 16) 100 | 25) 1,000 | 34) 85,000 |
| 8) 70 | 17) 300 | 26) 3,000 | 35) 71,000 |
| 9) 50 | 18) 400 | 27) 4,000 | 36) 27,000 |

**Rounding and Estimates**

| | | | |
|---|---|---|---|
| 1) 30 | 7) 760 | 13) 3,200 | 19) 130 |
| 2) 90 | 8) 4,610 | 14) 1,800 | 20) 110 |
| 3) 120 | 9) 2,100 | 15) 600 | 21) 400 |
| 4) 150 | 10) 300 | 16) 4,200 | 22) 100 |
| 5) 620 | 11) 2,400 | 17) 300 | |
| 6) 1,230 | 12) 200 | 18) 1,200 | |

**Adding and Subtracting Integers**

| | | | |
|---|---|---|---|
| 1) 8 | 8) 13 | 15) 42 | 22) 9 |
| 2) −33 | 9) 29 | 16) 36 | 23) 15 |
| 3) −23 | 10) 34 | 17) 101 | 24) 11 |
| 4) 38 | 11) −11 | 18) 40 | 25) 49 |
| 5) −19 | 12) 39 | 19) 32 | 26) 38 |
| 6) 24 | 13) −56 | 20) 112 | |
| 7) 33 | 14) 53 | 21) 40 | |

**Multiplying and Dividing Integers**

| | | | |
|---|---|---|---|
| 1) 45 | 6) −88 | 11) −6 | 16) 12 |
| 2) −27 | 7) 324 | 12) 8 | 17) −18 |
| 3) −96 | 8) −300 | 13) 5 | 18) 26 |
| 4) 140 | 9) −120 | 14) −27 | 19) −17 |
| 5) 90 | 10) −400 | 15) 8 | 20) −21 |

21) 23        23) 30        25) 26
22) 19        24) −15       26) −9

**Order of Operations**

1) 27         6) 24         11) 40        16) 56
2) −4         7) −16        12) 36        17) 5
3) 92         8) 50         13) −7        18) 74
4) −7         9) 43         14) 10        19) 60
5) 33         10) 27        15) 576       20) −47

**Ordering Integers and Numbers**

1) −10, −5, −3, 4, 8          7) 36, 18, −10, −16, −18
2) −18, −10, 6, 14, 27        8) 94, 34, 27, −12, −24
3) −23, −21, −8, 15, 21       9) 50, 42, −2, −13, −21
4) −40, −14, −12, 23, 47      10) 86, 46, 37, −16, −20
5) −57, −54, 32, 36, 59       11) 88, 75, −18, −26, −59
6) −34, −19, 26, 47, 68       12) 14, 3, −25, −30, −65

**Integers and Absolute Value**

1) 2          8) 2          15) 32        22) −2
2) 27         9) 54         16) 17        23) 32
3) 20         10) 4         17) 42        24) 19
4) 14         11) 11        18) 46        25) 28
5) 6          12) 88        19) 1         26) 10
6) 55         13) 0         20) 40        27) 28
7) 16         14) 79        21) 2         28) 88

**Factoring Numbers**

1) 1, 3, 9                    9) 1, 2, 4, 7, 14, 28         17) 1, 2, 5, 10, 25, 50
2) 1, 2, 4, 8, 16            10) 1, 2, 7, 14, 49, 98        18) 1, 2, 31, 62
3) 1, 2, 3, 4, 6, 8, 12, 24  11) 1, 2, 7, 14                19) 1, 5, 19, 95
4) 1, 2, 3, 5, 6, 10, 15, 30 12) 1, 2, 3, 6, 9, 18, 27, 54  20) 1, 2, 4, 8, 16, 32, 64
5) 1, 2, 13, 26             13) 1, 5, 11, 55               21) 1, 2, 5, 7, 10, 14, 35, 70
6) 1, 2, 23, 46            14) 1, 2, 3, 6, 9, 18           22) 1, 3, 5, 9, 15, 45
7) 1, 2, 4, 5, 10, 20      15) 1, 3, 7, 9, 21, 63          23) 1, 2, 11, 22
8) 1, 2, 4, 17, 34, 68     16) 1, 2, 17, 34                24) 1, 5, 13, 65

**Greatest Common Factor**

| | | | |
|---|---|---|---|
| 1) 2 | 7) 6 | 13) 5 | 19) 1 |
| 2) 1 | 8) 3 | 14) 2 | 20) 5 |
| 3) 3 | 9) 6 | 15) 9 | 21) 1 |
| 4) 1 | 10) 4 | 16) 2 | 22) 16 |
| 5) 5 | 11) 2 | 17) 14 | 23) 6 |
| 6) 8 | 12) 4 | 18) 8 | 24) 9 |

**Least Common Multiple**

| | | | |
|---|---|---|---|
| 1) 18 | 7) 30 | 13) 44 | 19) 180 |
| 2) 45 | 8) 56 | 14) 56 | 20) 105 |
| 3) 80 | 9) 21 | 15) 56 | 21) 162 |
| 4) 36 | 10) 40 | 16) 200 | 22) 270 |
| 5) 54 | 11) 75 | 17) 156 | 23) 90 |
| 6) 42 | 12) 63 | 18) 22 | 24) 75 |

# Chapter 2 :

# Fractions and Decimals

## Topics that you'll practice in this chapter:

- ✓ Simplifying Fractions
- ✓ Adding and Subtracting Fractions
- ✓ Multiplying and Dividing Fractions
- ✓ Adding and Subtract Mixed Numbers
- ✓ Multiplying and Dividing Mixed Numbers
- ✓ Adding and Subtracting Decimals
- ✓ Multiplying and Dividing Decimals
- ✓ Comparing Decimals
- ✓ Rounding Decimals

*"A Man is like a fraction whose numerator is what he is and whose denominator is what he thinks of himself. The larger the denominator, the smaller the fraction." –Tolstoy*

# Simplifying Fractions

✎ **Simplify each fraction to its lowest terms.**

1) $\dfrac{5}{10} =$

2) $\dfrac{28}{35} =$

3) $\dfrac{27}{36} =$

4) $\dfrac{40}{80} =$

5) $\dfrac{14}{56} =$

6) $\dfrac{32}{48} =$

7) $\dfrac{52}{65} =$

8) $\dfrac{15}{60} =$

9) $\dfrac{80}{160} =$

10) $\dfrac{55}{77} =$

11) $\dfrac{28}{112} =$

12) $\dfrac{32}{64} =$

13) $\dfrac{63}{72} =$

14) $\dfrac{81}{90} =$

15) $\dfrac{35}{105} =$

16) $\dfrac{25}{70} =$

17) $\dfrac{80}{280} =$

18) $\dfrac{12}{81} =$

19) $\dfrac{36}{186} =$

20) $\dfrac{240}{540} =$

21) $\dfrac{70}{560} =$

✎ **Find the answer for each problem.**

22) Which of the following fractions equal to $\dfrac{3}{4}$? ____

A. $\dfrac{60}{90}$       B. $\dfrac{43}{104}$       C. $\dfrac{48}{64}$       D. $\dfrac{150}{300}$

23) Which of the following fractions equal to $\dfrac{5}{8}$? ____

A. $\dfrac{125}{200}$       B. $\dfrac{115}{200}$       C. $\dfrac{50}{100}$       D. $\dfrac{30}{90}$

24) Which of the following fractions equal to $\dfrac{3}{7}$? ____

A. $\dfrac{58}{116}$       B. $\dfrac{54}{126}$       C. $\dfrac{270}{167}$       D. $\dfrac{42}{63}$

# Adding and Subtracting Fractions

✎ **Find the sum.**

1) $\frac{5}{9} + \frac{4}{9} =$

5) $\frac{1}{4} + \frac{3}{5} =$

9) $\frac{5}{7} + \frac{2}{3} =$

2) $\frac{1}{2} + \frac{1}{7} =$

6) $\frac{7}{8} + \frac{3}{8} =$

10) $\frac{7}{12} + \frac{3}{4} =$

3) $\frac{3}{8} + \frac{1}{4} =$

7) $\frac{1}{2} + \frac{7}{10} =$

11) $\frac{5}{6} + \frac{2}{5} =$

4) $\frac{3}{5} + \frac{1}{2} =$

8) $\frac{2}{5} + \frac{2}{3} =$

12) $\frac{1}{12} + \frac{2}{3} =$

✎ **Find the difference.**

13) $\frac{1}{3} - \frac{1}{6} =$

19) $\frac{5}{6} - \frac{1}{9} =$

25) $\frac{6}{7} - \frac{3}{4} =$

14) $\frac{3}{4} - \frac{1}{8} =$

20) $\frac{3}{4} - \frac{1}{6} =$

26) $\frac{4}{5} - \frac{1}{8} =$

15) $\frac{1}{2} - \frac{1}{3} =$

21) $\frac{7}{8} - \frac{1}{12} =$

27) $\frac{4}{7} - \frac{2}{35} =$

16) $\frac{1}{4} - \frac{1}{5} =$

22) $\frac{8}{15} - \frac{3}{5} =$

28) $\frac{9}{16} - \frac{2}{8} =$

17) $\frac{5}{8} - \frac{2}{3} =$

23) $\frac{3}{12} - \frac{1}{14} =$

29) $\frac{8}{9} - \frac{7}{18} =$

18) $\frac{1}{4} - \frac{1}{7} =$

24) $\frac{10}{13} - \frac{7}{26} =$

30) $\frac{1}{2} - \frac{4}{9} =$

# Multiplying and Dividing Fractions

✎ Find the value of each expression in lowest terms.

1) $\frac{1}{5} \times \frac{15}{5} =$

2) $\frac{9}{12} \times \frac{4}{9} =$

3) $\frac{1}{16} \times \frac{8}{10} =$

4) $\frac{1}{24} \times \frac{8}{10} =$

5) $\frac{1}{5} \times \frac{1}{4} =$

6) $\frac{7}{9} \times \frac{1}{7} =$

7) $\frac{6}{7} \times \frac{1}{3} =$

8) $\frac{2}{8} \times \frac{2}{8} =$

9) $\frac{5}{8} \times \frac{3}{5} =$

10) $\frac{4}{7} \times \frac{1}{8} =$

11) $\frac{7}{15} \times \frac{5}{7} =$

12) $\frac{3}{10} \times \frac{5}{9} =$

✎ Find the value of each expression in lowest terms.

13) $\frac{1}{4} \div \frac{1}{8} =$

14) $\frac{1}{10} \div \frac{1}{5} =$

15) $\frac{3}{4} \div \frac{1}{5} =$

16) $\frac{1}{3} \div \frac{5}{6} =$

17) $\frac{1}{7} \div \frac{8}{42} =$

18) $\frac{3}{4} \div \frac{1}{6} =$

19) $\frac{2}{7} \div \frac{7}{13} =$

20) $\frac{1}{24} \div \frac{3}{16} =$

21) $\frac{7}{12} \div \frac{5}{6} =$

22) $\frac{22}{18} \div \frac{11}{9} =$

23) $\frac{9}{35} \div \frac{3}{7} =$

24) $\frac{2}{7} \div \frac{8}{21} =$

25) $\frac{1}{9} \div \frac{2}{5} =$

26) $\frac{5}{12} \div \frac{3}{5} =$

27) $\frac{3}{20} \div \frac{1}{6} =$

28) $\frac{8}{20} \div \frac{3}{4} =$

29) $\frac{5}{6} \div \frac{2}{9} =$

30) $\frac{5}{11} \div \frac{3}{4} =$

## Adding and Subtracting Mixed Numbers

✎ Find the sum.

1) $3\frac{1}{3} + 2\frac{1}{6} =$

2) $4\frac{1}{2} + 3\frac{1}{2} =$

3) $3\frac{3}{8} + 1\frac{1}{8} =$

4) $2\frac{1}{4} + 2\frac{1}{3} =$

5) $3\frac{5}{6} + 2\frac{7}{12} =$

6) $5\frac{4}{15} + 3\frac{3}{5} =$

7) $2\frac{1}{3} + 4\frac{3}{7} =$

8) $3\frac{1}{2} + 4\frac{2}{5} =$

9) $5\frac{2}{5} + 6\frac{3}{7} =$

10) $8\frac{5}{16} + 6\frac{1}{12} =$

✎ Find the difference.

11) $3\frac{1}{4} - 1\frac{3}{4} =$

12) $6\frac{3}{5} - 4\frac{2}{5} =$

13) $4\frac{1}{3} - 3\frac{1}{9} =$

14) $7\frac{1}{7} - 5\frac{1}{2} =$

15) $5\frac{1}{3} - 2\frac{1}{12} =$

16) $8\frac{1}{5} - 4\frac{1}{3} =$

17) $9\frac{1}{4} - 6\frac{1}{8} =$

18) $11\frac{7}{15} - 8\frac{3}{5} =$

19) $14\frac{5}{6} - 11\frac{3}{5} =$

20) $18\frac{2}{7} - 14\frac{1}{5} =$

21) $9\frac{1}{3} - 4\frac{1}{4} =$

22) $6\frac{1}{8} - 4\frac{1}{16} =$

23) $19\frac{3}{8} - 15\frac{1}{3} =$

24) $11\frac{1}{9} - 8\frac{1}{8} =$

25) $17\frac{1}{7} - 11\frac{1}{5} =$

26) $16\frac{2}{9} - 9\frac{5}{7} =$

# Multiplying and Dividing Mixed Numbers

✍ **Find the product.**

1) $5\frac{1}{2} \times 2\frac{1}{4} =$

2) $5\frac{1}{3} \times 4\frac{1}{3} =$

3) $5\frac{3}{4} \times 6\frac{1}{4} =$

4) $3\frac{1}{3} \times 2\frac{3}{5} =$

5) $4\frac{8}{10} \times 1\frac{1}{24} =$

6) $6\frac{2}{7} \times 1\frac{1}{11} =$

7) $8\frac{2}{3} \times 3\frac{1}{2} =$

8) $3\frac{4}{7} \times 2\frac{1}{5} =$

9) $5\frac{2}{8} \times 4\frac{1}{6} =$

10) $7\frac{3}{3} \times 1\frac{3}{8} =$

✍ **Find the quotient.**

11) $2\frac{2}{5} \div 4\frac{1}{5} =$

12) $4\frac{1}{6} \div 3\frac{1}{3} =$

13) $6\frac{1}{3} \div 1\frac{1}{2} =$

14) $7\frac{1}{10} \div 2\frac{2}{5} =$

15) $3\frac{1}{3} \div 1\frac{1}{9} =$

16) $1\frac{1}{10} \div 4\frac{1}{2} =$

17) $1\frac{3}{16} \div 5\frac{1}{4} =$

18) $4\frac{1}{3} \div 4\frac{3}{4} =$

19) $9\frac{1}{3} \div 2\frac{1}{4} =$

20) $15\frac{1}{3} \div 5\frac{1}{2} =$

21) $4\frac{1}{6} \div 1\frac{1}{5} =$

22) $1\frac{1}{18} \div 1\frac{2}{9} =$

23) $4\frac{2}{7} \div 1\frac{3}{10} =$

24) $7\frac{1}{3} \div 2\frac{2}{11} =$

25) $8\frac{2}{5} \div 1\frac{1}{6} =$

26) $9\frac{1}{3} \div 2\frac{1}{7} =$

# Adding and Subtracting Decimals

✎ **Add and subtract decimals.**

1) $\begin{array}{r} 35.19 \\ -\ 24.28 \\ \hline \end{array}$

4) $\begin{array}{r} 38.72 \\ -\ 21.68 \\ \hline \end{array}$

7) $\begin{array}{r} 86.09 \\ -\ 35.14 \\ \hline \end{array}$

2) $\begin{array}{r} 34.29 \\ +\ 42.58 \\ \hline \end{array}$

5) $\begin{array}{r} 57.39 \\ +\ 26.54 \\ \hline \end{array}$

8) $\begin{array}{r} 54.51 \\ +\ 32.66 \\ \hline \end{array}$

3) $\begin{array}{r} 61.20 \\ +\ 33.75 \\ \hline \end{array}$

6) $\begin{array}{r} 70.24 \\ -\ 42.35 \\ \hline \end{array}$

9) $\begin{array}{r} 114.21 \\ -\ 88.69 \\ \hline \end{array}$

✎ **Find the missing number.**

10) ___ $+ 2.8 = 5.4$

15) ___ $- 19.85 = 6.54$

11) $4.1 +$ ___ $= 5.88$

16) $22.15 +$ ___ $= 28.95$

12) $6.45 +$ ___ $= 8$

17) ___ $- 37.16 = 9.42$

13) $7.25 -$ ___ $= 3.40$

18) ___ $+ 24.50 = 34.19$

14) ___ $- 2.35 = 4.25$

19) $72.40 +$ ___ $= 125.20$

# Multiplying and Dividing Decimals

✎ **Find the product.**

1) $0.5 \times 0.6 =$                     7) $5.25 \times 1.4 =$

2) $3.3 \times 0.4 =$                     8) $18.5 \times 4.6 =$

3) $1.28 \times 0.5 =$                    9) $15.4 \times 6.8 =$

4) $0.35 \times 0.6 =$                    10) $19.5 \times 2.6 =$

5) $1.85 \times 0.6 =$                    11) $32.2 \times 1.5 =$

6) $0.24 \times 0.5 =$                    12) $78.4 \times 4.5 =$

✎ **Find the quotient.**

13) $1.85 \div 10 =$                      19) $22.15 \div 1,000 =$

14) $74.6 \div 100 =$                     20) $53.55 \div 0.7 =$

15) $3.6 \div 3 =$                        21) $322.2 \div 0.2 =$

16) $9.6 \div 0.4 =$                      22) $50.67 \div 0.18 =$

17) $15.5 \div 0.5 =$                     23) $77.4 \div 0.8 =$

18) $32.8 \div 0.2 =$                     24) $27.93 \div 0.03 =$

## Comparing Decimals

✎ **Write the correct comparison symbol (>, < or =).**

1) 0.70 ☐ 0.070

2) 0.049 ☐ 0.49

3) 5.090 ☐ 5.09

4) 2.57 ☐ 2.05

5) 9.03 ☐ 0.930

6) 6.06 ☐ 6.6

7) 7.02 ☐ 7.020

8) 3.04 ☐ 3.2

9) 3.61 ☐ 3.245

10) 0.986 ☐ 0.0986

11) 17.24 ☐ 17.240

12) 0.759 ☐ 0.81

13) 9.040 ☐ 9.40

14) 5.73 ☐ 5.213

15) 9.44 ☐ 9.404

16) 7.17 ☐ 7.170

17) 4.85 ☐ 4.085

18) 9.041 ☐ 9.40

19) 3.033 ☐ 3.030

20) 4.97 ☐ 4.970

# Rounding Decimals

✍ **Round each decimal to the nearest whole number.**

1) 28.12

3) 16.22

5) 7.95

2) 6.9

4) 8.5

6) 52.7

✍ **Round each decimal to the nearest tenth.**

7) 31.761

9) 94.729

11)  13.219

8) 14.421

10)  77.89

12)  59.89

✍ **Round each decimal to the nearest hundredth.**

13) 8.428

15) 55.3786

17) 62.241

14) 23.812

16) 231.912

18) 19.447

✍ **Round each decimal to the nearest thousandth.**

19) 15.54324

21) 243.8652

23) 67.1983

20) 34.62586

22) 80.4529

24) 72.36788

# Answers of Worksheets

**Simplifying Fractions**

1) $\frac{1}{2}$

2) $\frac{4}{5}$

3) $\frac{3}{4}$

4) $\frac{1}{2}$

5) $\frac{1}{4}$

6) $\frac{2}{3}$

7) $\frac{4}{5}$

8) $\frac{1}{4}$

9) $\frac{1}{2}$

10) $\frac{5}{7}$

11) $\frac{1}{4}$

12) $\frac{1}{2}$

13) $\frac{7}{8}$

14) $\frac{9}{10}$

15) $\frac{1}{3}$

16) $\frac{5}{14}$

17) $\frac{2}{7}$

18) $\frac{4}{27}$

19) $\frac{6}{31}$

20) $\frac{4}{9}$

21) $\frac{1}{8}$

22) C

23) A

24) B

**Adding and Subtracting Fractions**

1) $\frac{9}{9} = 1$

2) $\frac{9}{14}$

3) $\frac{5}{8}$

4) $1\frac{1}{10}$

5) $\frac{17}{20}$

6) $1\frac{1}{4}$

7) $1\frac{1}{5}$

8) $1\frac{1}{15}$

9) $1\frac{8}{21}$

10) $1\frac{1}{3}$

11) $1\frac{7}{30}$

12) $\frac{3}{4}$

13) $\frac{1}{6}$

14) $\frac{5}{8}$

15) $\frac{1}{6}$

16) $\frac{1}{20}$

17) $-\frac{1}{24}$

18) $\frac{3}{28}$

19) $\frac{13}{18}$

20) $\frac{7}{12}$

21) $\frac{19}{24}$

22) $-\frac{1}{15}$

23) $\frac{5}{28}$

24) $\frac{1}{2}$

25) $\frac{3}{28}$

26) $\frac{27}{40}$

27) $\frac{18}{35}$

28) $\frac{5}{16}$

29) $\frac{1}{2}$

30) $\frac{1}{18}$

**Multiplying and Dividing Fractions**

1) $\frac{3}{5}$

2) $\frac{1}{3}$

3) $\frac{1}{20}$

4) $\frac{1}{30}$

5) $\frac{1}{20}$

6) $\frac{1}{9}$

7) $\frac{2}{7}$

8) $\frac{1}{16}$

9) $\frac{3}{8}$

10) $\frac{1}{14}$

11) $\frac{1}{3}$

12) $\frac{1}{6}$

13) 2

14) $\frac{1}{2}$

15) $3\frac{3}{4}$

16) $\frac{2}{5}$

17) $\frac{3}{4}$

18) $4\frac{1}{2}$

19) $\frac{26}{49}$

20) $\frac{2}{9}$

21) $\frac{7}{10}$

22) $1$

23) $\frac{3}{5}$

24) $\frac{3}{4}$

25) $\frac{5}{18}$

26) $\frac{25}{36}$

27) $\frac{9}{10}$

28) $\frac{8}{15}$

29) $3\frac{3}{4}$

30) $\frac{20}{33}$

## Adding and Subtracting Mixed Numbers

1) $5\frac{1}{2}$

2) $8$

3) $4\frac{1}{2}$

4) $4\frac{7}{12}$

5) $6\frac{5}{12}$

6) $8\frac{13}{15}$

7) $6\frac{16}{21}$

8) $7\frac{9}{10}$

9) $11\frac{29}{35}$

10) $14\frac{19}{48}$

11) $1\frac{1}{2}$

12) $2\frac{1}{5}$

13) $1\frac{2}{9}$

14) $1\frac{9}{14}$

15) $3\frac{1}{4}$

16) $3\frac{13}{15}$

17) $3\frac{1}{8}$

18) $2\frac{13}{15}$

19) $3\frac{7}{30}$

20) $4\frac{3}{35}$

21) $5\frac{1}{12}$

22) $2\frac{1}{16}$

23) $4\frac{1}{24}$

24) $2\frac{71}{72}$

25) $5\frac{33}{35}$

26) $6\frac{32}{63}$

## Multiplying and Dividing Mixed Numbers

1) $12\frac{3}{8}$

2) $23\frac{1}{9}$

3) $35\frac{15}{16}$

4) $8\frac{2}{3}$

5) $5$

6) $6\frac{6}{7}$

7) $30\frac{1}{3}$

8) $7\frac{6}{7}$

9) $21\frac{7}{8}$

10) $11$

11) $\frac{4}{7}$

12) $1\frac{1}{4}$

13) $4\frac{2}{9}$

14) $2\frac{23}{24}$

15) $3$

16) $\frac{11}{45}$

17) $\frac{19}{84}$

18) $\frac{52}{57}$

19) $4\frac{4}{27}$

20) $2\frac{26}{33}$

21) $3\frac{17}{36}$

22) $\frac{19}{22}$

23) $3\frac{27}{91}$

24) $3\frac{13}{36}$

25) $7\frac{1}{5}$

26) $4\frac{16}{45}$

## Adding and Subtracting Decimals

1) 10.91

2) 76.87

3) 94.95

4) 17.04

| | | | |
|---|---|---|---|
| 5) 83.93 | 9) 25.52 | 13) 3.85 | 17) 46.58 |
| 6) 27.89 | 10) 2.6 | 14) 6.6 | 18) 9.69 |
| 7) 50.95 | 11) 1.78 | 15) 26.39 | 19) 52.8 |
| 8) 87.17 | 12) 1.55 | 16) 6.8 | |

**Multiplying and Dividing Decimals**

| | | | |
|---|---|---|---|
| 1) 0.3 | 7) 7.35 | 13) 0.185 | 19) 0.02215 |
| 2) 1.32 | 8) 85.1 | 14) 0.746 | 20) 76.5 |
| 3) 0.64 | 9) 104.72 | 15) 1.2 | 21) 1,611 |
| 4) 0.21 | 10) 50.7 | 16) 24 | 22) 281.5 |
| 5) 1.11 | 11) 48.3 | 17) 31 | 23) 96.75 |
| 6) 0.12 | 12) 352.8 | 18) 164 | 24) 931 |

**Comparing Decimals**

| | | | |
|---|---|---|---|
| 1) > | 6) < | 11) = | 16) = |
| 2) < | 7) = | 12) < | 17) > |
| 3) = | 8) < | 13) < | 18) < |
| 4) > | 9) > | 14) > | 19) > |
| 5) > | 10) > | 15) > | 20) = |

**Rounding Decimals**

| | | |
|---|---|---|
| 1) 28 | 9) 94.7 | 17) 62.24 |
| 2) 7 | 10) 77.9 | 18) 19.45 |
| 3) 16 | 11) 13.2 | 19) 15.543 |
| 4) 9 | 12) 59.9 | 20) 34.626 |
| 5) 8 | 13) 8.43 | 21) 243.865 |
| 6) 53 | 14) 23.81 | 22) 80.453 |
| 7) 31.8 | 15) 55.38 | 23) 67.198 |
| 8) 14.4 | 16) 231.91 | 24) 72.368 |

# Chapter 3 :

# Proportions, Ratios, and Percent

## Topics that you'll practice in this chapter:

- ✓ Simplifying Ratios
- ✓ Proportional Ratios
- ✓ Similarity and Ratios
- ✓ Ratio and Rates Word Problems
- ✓ Percentage Calculations
- ✓ Percent Problems
- ✓ Discount, Tax and Tip
- ✓ Percent of Change
- ✓ Simple Interest

*Without mathematics, there's nothing you can do. Everything around you is mathematics.*
*Everything around you is numbers." – Shakuntala Devi*

# Simplifying Ratios

✎ **Reduce each ratio.**

1) $15:20 = $ ___ : ___   9) $8:48 = $ ___ : ___   17) $56:72 = $ ___ : ___

2) $7:70 = $ ___ : ___   10) $49:63 = $ ___ : ___   18) $26:13 = $ ___ : ___

3) $16:28 = $ ___ : ___   11) $18:27 = $ ___ : ___   19) $15:45 = $ ___ : ___

4) $7:21 = $ ___ : ___   12) $35:10 = $ ___ : ___   20) $28:4 = $ ___ : ___

5) $4:40 = $ ___ : ___   13) $90:9 = $ ___ : ___   21) $24:48 = $ ___ : ___

6) $6:48 = $ ___ : ___   14) $24:32 = $ ___ : ___   22) $30:24 = $ ___ : ___

7) $16:64 = $ ___ : ___   15) $7:56 = $ ___ : ___   23) $70:140 = $ ___ : ___

8) $10:25 = $ ___ : ___   16) $45:63 = $ ___ : ___   24) $6:180 = $ ___ : ___

✎ **Write each ratio as a fraction in simplest form.**

25) $6:12 = $   32) $7:35 = $   39) $20:300 = $

26) $30:50 = $   33) $40:96 = $   40) $30:120 = $

27) $15:35 = $   34) $12:54 = $   41) $56:42 = $

28) $9:27 = $   35) $44:52 = $   42) $26:130 = $

29) $8:24 = $   36) $12:27 = $   43) $66:123 = $

30) $18:84 = $   37) $15:180 = $   44) $70:630 = $

31) $7:14 = $   38) $39:143 = $   45) $75:125 = $

# Proportional Ratios

✎ Fill in the blanks; Calculate each proportion.

1) $3 : 8 = \underline{\quad} : 48$

2) $2 : 5 = 20 : \underline{\quad}$

3) $1 : 9 = \underline{\quad} : 81$

4) $6 : 7 = 12 : \underline{\quad}$

5) $9 : 2 = 63 : \underline{\quad}$

6) $8 : 7 = \underline{\quad} : 49$

7) $20 : 3 = \underline{\quad} : 15$

8) $1 : 3 = \underline{\quad} : 75$

9) $7 : 6 = \underline{\quad} : 60$

10) $8 : 5 = \underline{\quad} : 45$

11) $3 : 10 = 60 : \underline{\quad}$

12) $6 : 11 = 42 : \underline{\quad}$

✎ State if each pair of ratios form a proportion.

13) $\frac{3}{20}$ and $\frac{9}{60}$

14) $\frac{1}{7}$ and $\frac{6}{42}$

15) $\frac{3}{7}$ and $\frac{24}{56}$

16) $\frac{4}{9}$ and $\frac{12}{18}$

17) $\frac{1}{9}$ and $\frac{12}{81}$

18) $\frac{7}{8}$ and $\frac{21}{28}$

19) $\frac{9}{13}$ and $\frac{27}{39}$

20) $\frac{1}{8}$ and $\frac{8}{64}$

21) $\frac{6}{19}$ and $\frac{30}{85}$

22) $\frac{5}{9}$ and $\frac{40}{81}$

23) $\frac{9}{14}$ and $\frac{108}{168}$

24) $\frac{15}{23}$ and $\frac{360}{552}$

✎ Calculate each proportion.

25) $\frac{20}{25} = \frac{32}{x}, x = \underline{\quad}$

26) $\frac{1}{8} = \frac{32}{x}, x = \underline{\quad}$

27) $\frac{15}{5} = \frac{21}{x}, x = \underline{\quad}$

28) $\frac{1}{7} = \frac{x}{294}, x = \underline{\quad}$

29) $\frac{7}{9} = \frac{x}{81}, x = \underline{\quad}$

30) $\frac{1}{5} = \frac{13}{x}, x = \underline{\quad}$

31) $\frac{9}{5} = \frac{36}{x}, x = \underline{\quad}$

32) $\frac{6}{13} = \frac{48}{x}, x = \underline{\quad}$

33) $\frac{5}{8} = \frac{x}{88}, x = \underline{\quad}$

34) $\frac{4}{15} = \frac{x}{240}, x = \underline{\quad}$

35) $\frac{9}{19} = \frac{x}{266}, x = \underline{\quad}$

36) $\frac{7}{15} = \frac{x}{270}, x = \underline{\quad}$

## Similarity and Ratios

✎ **Each pair of figures is similar. Find the missing side.**

1)
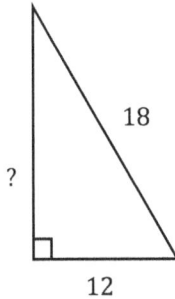

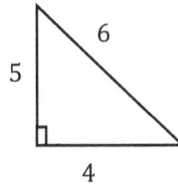

2)
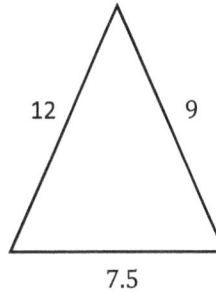

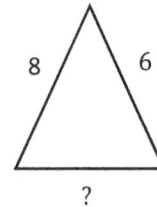

3)
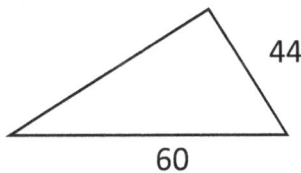

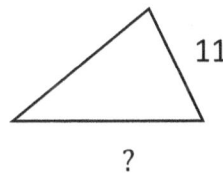

4)
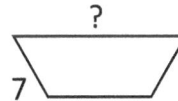

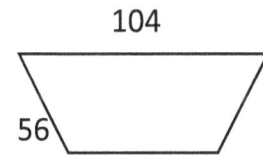

✎ **Calculate.**

5) Two rectangles are similar. The first is 24 feet wide and 120 feet long. The second is 30 feet wide. What is the length of the second rectangle? _____

6) Two rectangles are similar. One is 5 meters by 36 meters. The longer side of the second rectangle is 90 meters. What is the other side of the second rectangle? _____

7) A building casts a shadow 25 ft long. At the same time a girl 10 ft tall casts a shadow 5 ft long. How tall is the building? _____

8) The scale of a map of Texas is 4 inches: 32 miles. If you measure the distance from Dallas to Martin County as 38.4 inches, approximately how far is Martin County from Dallas? _____

# Ratio and Rates Word Problems

✎ **Find the answer for each word problem.**

1) Mason has 24 red cards and 36 green cards. What is the ratio of Mason 's red cards to his green cards? _____

2) In a party, 45 soft drinks are required for every 54 guests. If there are 378 guests, how many soft drinks is required? _____

3) In Mason's class, 42 of the students are tall and 24 are short. In Michael's class 84 students are tall and 48 students are short. Which class has a higher ratio of tall to short students? _____

4) The price of 5 apples at the Quick Market is $4.6. The price of 7 of the same apples at Walmart is $5.95. Which place is the better buy? _____

5) The bakers at a Bakery can make 90 bagels in 3 hours. How many bagels can they bake in 24 hours? What is that rate per hour? _____

6) You can buy 5 cans of green beans at a supermarket for $5.75. How much does it cost to buy 45 cans of green beans? _____

7) The ratio of boys to girls in a class is 4: 7. If there are 32 boys in the class, how many girls are in that class? _____

8) The ratio of red marbles to blue marbles in a bag is 3: 7. If there are 50 marbles in the bag, how many of the marbles are red? _____

# Percentage Calculations

✎ **Calculate the given percent of each value.**

1) 3% of 60 = ____

2) 20% of 32 = ____

3) 4% of 72 = ____

4) 16% of 32 = ____

5) 25% of 124 = ____

6) 35% of 56 = ____

7) 15% of 20 = ____

8) 14% of 150 = ____

9) 80% of 50 = ____

10) 12% of 115 = ____

11) 72% of 250 = ____

12) 52% of 500 = ____

13) 70% of 400 = ____

14) 27% of 145 = ____

15) 90% of 64 = ____

16) 60% of 55 = ____

17) 22% of 210 = ____

18) 8% of 235 = ____

✎ **Calculate the percent of each given value.**

19) ____% of 25 = 5

20) ____% of 40 = 20

21) ____% of 25 = 2

22) ____% of 50 = 16

23) ____% of 250 = 5

24) ____% of 40 = 32

25) ____% of 125 = 20

26) ____% of 700 = 49

27) ____% of 350 = 49

28) ____% of 500 = 210

✎ **Calculate each percent problem.**

29) A Cinema has 250 seats. 60 seats were sold for the current movie. What percent of seats are empty? _____ %

30) There are 68 boys and 92 girls in a class. 75% of the students in the class take the bus to school. How many students do not take the bus to school? _____

# Percent Problems

✍ **Calculate each problem.**

1) 9 is what percent of 45? ____%

2) 60 is what percent of 120? ____%

3) 10 is what percent of 200? ____%

4) 15 is what percent of 125? ____%

5) 10 is what percent of 400? ____%

6) 66 is what percent of 55? ____%

7) 40 is what percent of 160? ____%

8) 40 is what percent of 50? ____%

9) 120 is what percent of 800? ____%

10) 78 is what percent of 120? ___%

11) 36 is what percent of 144? ___%

12) 17 is what percent of 85? ___%

13) 90 is what percent of 900? ___%

14) 36 is what percent of 16? ___%

15) 63 is what percent of 14? ___%

16) 18 is what percent of 60? ___%

17) 126 is what percent of 200? ___%

18) 232 is what percent of 40? ___%

✍ **Calculate each percent word problem.**

19) There are 40 employees in a company. On a certain day, 25 were present. What percent showed up for work? ____%

20) A metal bar weighs 60 ounces. 25% of the bar is gold. How many ounces of gold are in the bar? _____

21) A crew is made up of 12 women; the rest are men. If 15% of the crew are women, how many people are in the crew? _____

22) There are 40 students in a class and 8 of them are girls. What percent are boys? ____%

23) The Royals softball team played 400 games and won 280 of them. What percent of the games did they lose? ____%

# Discount, Tax and Tip

## Find the selling price of each item.

1) Original price of a computer: $420

   Tax: 8% Selling price: $_____

2) Original price of a laptop: $280

   Tax: 4% Selling price: $_____

3) Original price of a sofa: $820

   Tax: 5% Selling price: $_____

4) Original price of a car: $15,800

   Tax: 3.6%   Selling price: $_____

5) Original price of a Table: $250

   Tax: 9% Selling price: $_____

6) Original price of a house: $630,000

   Tax: 1.8%   Selling price: $_____

7) Original price of a tablet: $450

   Discount: 30%   Selling price: $____

8) Original price of a chair: $390

   Discount: 8%    Selling price: $____

9) Original price of a book: $75

   Discount: 42%   Selling price: $____

10) Original price of a cellphone: $820

    Discount: 23%   Selling price: $___

11) Food bill: $45

    Tip: 15%         Price: $_____

12) Food bill: $32

    Tipp: 20%        Price: $_____

13) Food bill: $90

    Tip: 35%         Price: $_____

14) Food bill: $42

    Tipp: 12%   Price: $_____

## Find the answer for each word problem.

15) Nicolas hired a moving company. The company charged $500 for its services, and Nicolas gives the movers a 40% tip. How much does Nicolas tip the movers? $_____

16) Mason has lunch at a restaurant and the cost of his meal is $90. Mason wants to leave a 25% tip. What is Mason's total bill including tip? $_____

17) The sales tax in Texas is 19.80% and an item costs $350. How much is the tax? $_____

18) The price of a table at Best Buy is $680. If the sales tax is 5%, what is the final price of the table including tax? $_____

# Percent of Change

### ✍ Find each percent of change.

1) From 150 to 450. ___ %

2) From 50 ft to 250 ft. ___ %

3) From $60 to $360. ___ %

4) From 60 cm to 180 cm. ___ %

5) From 15 to 45. ___ %

6) From 80 to 16. ___ %

7) From 120 to 360. ___ %

8) From 900 to 450. ___ %

9) From 1,000 to 200. ___ %

10) From 144 to 36. ___ %

### ✍ Calculate each percent of change word problem.

11) Bob got a raise, and his hourly wage increased from $42 to $63. What is the percent increase? ___ %

12) The price of a pair of shoes increases from $50 to $61. What is the percent increase? ___ %

13) At a coffee shop, the price of a cup of coffee increased from $4.80 to $5.76. What is the percent increase in the cost of the coffee? ___ %

14) 51 cm are cut from 85 cm board. What is the percent decrease in length? ___ %

15) In a class, the number of students has been increased from 54 to 81. What is the percent increase? ___ %

16) The price of gasoline rises from $24.40 to $30.50 in one month. By what percent did the gas price rise? ___ %

17) A shirt was originally priced at $38. It went on sale for $24.70. What was the percent that the shirt was discounted? ___ %

# Simple Interest

## ✍ Determine the simple interest for these loans.

1) $480 at 11% for 3 years. $ _____

2) $4,200 at 7% for 4 years. $ _____

3) $2,500 at 20% for 3 years. $ _____

4) $6,800 at 3.9% for 4 months. $ ____

5) $800 at 6% for 7 months. $ _____

6) $36,000 at 4.2% for 6 years. $ ____

7) $6,500 at 7% for 4 years. $ _____

8) $850 at 9.5% for 2 years. $ _____

9) $1,200 at 5.8% for 9 months. $ ____

10) $3,000 at 4.5% for 7 years. $ ____

## ✍ Calculate each simple interest word problem.

11) A new car, valued at $22,000, depreciates at 8.5% per year. What is the value of the car one year after purchase? $_____

12) Sara puts $9,000 into an investment yielding 6% annual simple interest; she left the money in for three years. How much interest does Sara get at the end of those three years? $_____

13) A bank is offering 12% simple interest on a savings account. If you deposit $16,400, how much interest will you earn in two years? $_____

14) $720 interest is earned on a principal of $6,000 at a simple interest rate of 4% interest per year. For how many years was the principal invested? _____

15) In how many years will $2,200 yield an interest of $440 at 4% simple interest? _____

16) Jim invested $8,000 in a bond at a yearly rate of 4.5%. He earned $1,440 in interest. How long was the money invested? _____

# Answers of Worksheets

**Simplifying Ratios**

| | | | |
|---|---|---|---|
| 1) 3 : 4 | 14) 3 : 4 | 26) $\frac{3}{5}$ | 36) $\frac{4}{9}$ |
| 2) 1 : 10 | 15) 1 : 8 | 27) $\frac{3}{7}$ | 37) $\frac{1}{12}$ |
| 3) 4 : 7 | 16) 5 : 7 | 28) $\frac{1}{3}$ | 38) $\frac{3}{11}$ |
| 4) 1 : 3 | 17) 7 : 9 | 29) $\frac{1}{3}$ | 39) $\frac{1}{15}$ |
| 5) 1 : 10 | 18) 2 : 1 | 30) $\frac{3}{14}$ | 40) $\frac{1}{4}$ |
| 6) 1 : 8 | 19) 1 : 3 | 31) $\frac{1}{2}$ | 41) $\frac{4}{3}$ |
| 7) 2 : 8 | 20) 7 : 1 | 32) $\frac{1}{5}$ | 42) $\frac{1}{5}$ |
| 8) 2 : 5 | 21) 1 : 2 | 33) $\frac{5}{12}$ | 43) $\frac{22}{41}$ |
| 9) 1 : 6 | 22) 5 : 4 | 34) $\frac{2}{9}$ | 44) $\frac{1}{9}$ |
| 10) 7 : 9 | 23) 1 : 2 | 35) $\frac{11}{13}$ | 45) $\frac{3}{5}$ |
| 11) 2 : 3 | 24) 1 : 30 | | |
| 12) 7 : 2 | 25) $\frac{1}{2}$ | | |
| 13) 10 : 1 | | | |

**Proportional Ratios**

| | | | |
|---|---|---|---|
| 1) 18 | 10) 72 | 19) Yes | 28) 42 |
| 2) 50 | 11) 200 | 20) Yes | 29) 63 |
| 3) 9 | 12) 77 | 21) No | 30) 65 |
| 4) 14 | 13) Yes | 22) No | 31) 20 |
| 5) 14 | 14) Yes | 23) Yes | 32) 104 |
| 6) 56 | 15) Yes | 24) Yes | 33) 55 |
| 7) 100 | 16) No | 25) 40 | 34) 64 |
| 8) 25 | 17) No | 26) 256 | 35) 126 |
| 9) 70 | 18) No | 27) 7 | 36) 126 |

**Similarity and ratios**

| | | |
|---|---|---|
| 1) 15 | 4) 13 | 7) 50 feet |
| 2) 5 | 5) 150 feet | 8) 307.2 miles |
| 3) 15 | 6) 12.5 meters | |

**Ratio and Rates Word Problems**

| | |
|---|---|
| 1) 2 : 3 | 2) 315 |

3) The ratio for both classes is 7 to 4.

4) Walmart is a better buy.

5) 720, the rate is 30 per hour.

6) $51.75

7) 56

8) 15

**Percentage Calculations**

| | | |
|---|---|---|
| 1) 1.8 | 11) 180 | 21) 8% |
| 2) 6.4 | 12) 260 | 22) 32% |
| 3) 2.88 | 13) 280 | 23) 2% |
| 4) 5.12 | 14) 39.15 | 24) 80% |
| 5) 31 | 15) 57.6 | 25) 16% |
| 6) 19.6 | 16) 33 | 26) 7% |
| 7) 3 | 17) 46.2 | 27) 14% |
| 8) 21 | 18) 18.8 | 28) 42% |
| 9) 40 | 19) 20% | 29) 76% |
| 10) 13.8 | 20) 50% | 30) 40 |

**Percent Problems**

| | | |
|---|---|---|
| 1) 20% | 9) 15% | 17) 63% |
| 2) 50% | 10) 65% | 18) 580% |
| 3) 5% | 11) 25% | 19) 62.5% |
| 4) 12% | 12) 20% | 20) 15 ounces |
| 5) 2.5% | 13) 10% | 21) 80 |
| 6) 120% | 14) 225% | 22) 80% |
| 7) 25% | 15) 450% | 23) 30% |
| 8) 80% | 16) 30% | |

**Discount, Tax and Tip**

| | | |
|---|---|---|
| 1) $453.60 | 7) $315.00 | 13) $121.50 |
| 2) $291.20 | 8) $358.80 | 14) $47.04 |
| 3) $861.00 | 9) $43.50 | 15) $200.00 |
| 4) $16,368.80 | 10) $631.40 | 16) $112.50 |
| 5) $272.50 | 11) $51.75 | 17) $69.30 |
| 6) $641,340 | 12) $38.40 | 18) $714.00 |

**Percent of Change**

| | | |
|---|---|---|
| 1) 200% | 7) 200% | 13) 20% |
| 2) 400% | 8) 50% | 14) 60% |
| 3) 500% | 9) 80% | 15) 50% |
| 4) 200% | 10) 75% | 16) 25% |
| 5) 200% | 11) 50% | 17) 35% |
| 6) 80% | 12) 22% | |

**Simple Interest**

| | | |
|---|---|---|
| 1) $158.40 | 7) $1,820.00 | 13) $3,936.00 |
| 2) $1,176.00 | 8) $161.50 | 14) 3 years |
| 3) $1,500.00 | 9) $52.20 | 15) 5 years |
| 4) $88.40 | 10) $945.00 | 16) 4 years |
| 5) $28.00 | 11) $20,130.00 | |
| 6) $9,072.00 | 12) $1,620.00 | |

# Chapter 4 :

# Exponents and Radicals Expressions

## Topics that you'll practice in this chapter:

- ✓ Multiplication Property of Exponents
- ✓ Zero and Negative Exponents
- ✓ Division Property of Exponents
- ✓ Powers of Products and Quotients
- ✓ Negative Exponents and Negative Bases
- ✓ Scientific Notation
- ✓ Square Roots
- ✓ Simplifying Radical Expressions
- ✓ Simplifying Radical Expressions Involving Fractions
- ✓ Multiplying Radical Expressions
- ✓ Adding and Subtracting Radical Expressions

*Love is anterior to life, posterior to death, initial of creation, and the exponent of breath.*

**Emily Dickinson**

# Multiplication Property of Exponents

✎ **Simplify and write the answer in exponential form.**

1) $4 \times 4^5 =$

2) $8^4 \times 8 =$

3) $7^3 \times 7^3 =$

4) $9^2 \times 9^2 =$

5) $2^2 \times 2^4 \times 2 =$

6) $5 \times 5^3 \times 5^3 =$

7) $4^3 \times 4^2 \times 4 \times 4 =$

8) $5x \times x =$

9) $x^3 \times x^3 =$

10) $x^7 \times x^2 =$

11) $x^4 \times x^3 \times x^2 =$

12) $10x \times 3x =$

13) $4x^3 \times 4x^3 =$

14) $7x^3 \times x =$

15) $3x^2 \times 4x^2 \times x^2 =$

16) $5x^4 \times x^4 =$

17) $2x^8 \times 2x =$

18) $6x \times x^5 =$

19) $4x^2 \times 6x^6 =$

20) $5yx^3 \times 4x =$

21) $7x^3 \times y^5 x^7 =$

22) $y^2 x^3 \times y^5 x^4 =$

23) $3x^5 \times 4x^3 y^4 =$

24) $4x^4 \times 9x^2 y^5 =$

25) $5x^3 y^4 \times 6x^8 y^2 =$

26) $8x^3 y^6 \times 4xy^3 =$

27) $2xy^5 \times 6x^3 y^3 =$

28) $4x^5 y^2 \times 4x^2 y^8 =$

29) $7x \times 3y^8 x^2 \times y^5 =$

30) $x^3 \times 2y^3 x^4 \times 2y =$

31) $3yx^4 \times 3y^4 x \times 3xy^3 =$

32) $6y^3 \times 2y^2 x^4 \times 10yx^5 =$

## Zero and Negative Exponents

✎ **Evaluate the following expressions.**

1) $1^{-5} =$

2) $4^{-1} =$

3) $0^{10} =$

4) $1^{15} =$

5) $5^{-2} =$

6) $3^{-3} =$

7) $9^{-1} =$

8) $10^{-2} =$

9) $12^{-2} =$

10) $2^{-5} =$

11) $3^{-4} =$

12) $2^{-4} =$

13) $6^{-3} =$

14) $10^{-3} =$

15) $30^{-1} =$

16) $15^{-2} =$

17) $4^{-3} =$

18) $2^{-7} =$

19) $5^{-3} =$

20) $4^{-4} =$

21) $3^{-5} =$

22) $10^{-4} =$

23) $2^{-10} =$

24) $8^{-3} =$

25) $20^{-2} =$

26) $14^{-2} =$

27) $9^{-3} =$

28) $100^{-2} =$

29) $5^{-4} =$

30) $4^{-6} =$

31) $\left(\frac{1}{4}\right)^{-3}$

32) $\left(\frac{1}{6}\right)^{-2} =$

33) $\left(\frac{1}{7}\right)^{-2} =$

34) $\left(\frac{2}{3}\right)^{-3} =$

35) $\left(\frac{1}{13}\right)^{-2} =$

36) $\left(\frac{7}{12}\right)^{-2} =$

37) $\left(\frac{1}{6}\right)^{-3} =$

38) $\left(\frac{1}{300}\right)^{-2} =$

39) $\left(\frac{2}{9}\right)^{-2} =$

40) $\left(\frac{7}{5}\right)^{-1} =$

41) $\left(\frac{13}{23}\right)^{0} =$

42) $\left(\frac{1}{4}\right)^{-5} =$

# Division Property of Exponents

✎ **Simplify.**

1) $\dfrac{5^6}{5^7} =$

2) $\dfrac{8^8}{8^6} =$

3) $\dfrac{4^5}{4} =$

4) $\dfrac{3}{3^5} =$

5) $\dfrac{x}{x^6} =$

6) $\dfrac{3 \times 3^2}{3^2 \times 3^5} =$

7) $\dfrac{9^4}{9^2} =$

8) $\dfrac{10 \times 10^9}{10^2 \times 10^7} =$

9) $\dfrac{7^5 \times 7^7}{7^4 \times 7^8} =$

10) $\dfrac{15x}{30x^6} =$

11) $\dfrac{3x^9}{4x^4} =$

12) $\dfrac{15x^8}{10x^9} =$

13) $\dfrac{42x^5}{6y^9} =$

14) $\dfrac{36y^8}{4x^4y^5} =$

15) $\dfrac{2x^7}{9x} =$

16) $\dfrac{49x^8y^6}{7x^9} =$

17) $\dfrac{48x^2}{24x^6y^{12}} =$

18) $\dfrac{30yx^5}{6yx^7} =$

19) $\dfrac{19x^7y}{38x^{12}y^4} =$

20) $\dfrac{9x^8}{63x^8} =$

21) $\dfrac{9x^{-9}}{4x^{-3}} =$

# Powers of Products and Quotients

✎ **Simplify.**

1) $(4^3)^2 =$

2) $(2^3)^4 =$

3) $(2 \times 2^3)^2 =$

4) $(5 \times 5^5)^6 =$

5) $(19^4 \times 19^2)^3 =$

6) $(2^3 \times 2^4)^4 =$

7) $(5 \times 5^2)^2 =$

8) $(4^4)^4 =$

9) $(8x^5)^2 =$

10) $(3x^2y^4)^4 =$

11) $(7x^5y^2)^2 =$

12) $(5x^4y^4)^3 =$

13) $(2x^3y^3)^5 =$

14) $(10x^3y^4)^3 =$

15) $(13y^3y)^2 =$

16) $(5x^6x^4)^2 =$

17) $(6x^7y^6)^3 =$

18) $(12x^5x^7)^2 =$

19) $(2x^4 \times 2x)^4 =$

20) $(2x^4y^3)^5 =$

21) $(15x^7y^2)^2 =$

22) $(8x^3y^5)^3 =$

23) $(3x \times 2y^2)^4 =$

24) $\left(\frac{4x}{x^5}\right)^2 =$

25) $\left(\frac{x^4y^5}{x^3y^5}\right)^9 =$

26) $\left(\frac{36xy}{6x^5}\right)^3 =$

27) $\left(\frac{x^7}{x^8y^2}\right)^6 =$

28) $\left(\frac{xy^4}{x^3y^6}\right)^{-3} =$

29) $\left(\frac{5xy^8}{x^3}\right)^2 =$

30) $\left(\frac{xy^6}{2xy^3}\right)^{-4} =$

## Negative Exponents and Negative Bases

✎ **Simplify.**

1) $-9^{-1} =$

2) $-9^{-2} =$

3) $-2^{-5} =$

4) $-x^{-7} =$

5) $11x^{-1} =$

6) $-8x^{-3} =$

7) $-12x^{-5} =$

8) $-9x^{-8}y^{-6} =$

9) $32x^{-5}y^{-1} =$

10) $10a^{-9}b^{-3} =$

11) $-17x^4y^{-6} =$

12) $-\dfrac{25}{x^{-5}} =$

13) $-\dfrac{13x}{a^{-7}} =$

14) $\left(-\dfrac{1}{3}\right)^{-4} =$

15) $\left(-\dfrac{3}{4}\right)^{-2} =$

16) $-\dfrac{14}{a^{-6}b^{-3}} =$

17) $-\dfrac{7x}{x^{-8}} =$

18) $-\dfrac{a^{-9}}{b^{-5}} =$

19) $-\dfrac{11}{x^{-5}} =$

20) $\dfrac{8b}{-16c^{-6}} =$

21) $\dfrac{12ab}{a^{-4}b^{-3}} =$

22) $-\dfrac{8n^{-4}}{32p^{-7}} =$

23) $\dfrac{16ab^{-6}}{-6c^{-5}} =$

24) $\left(\dfrac{10a}{5c}\right)^{-4} =$

25) $\left(-\dfrac{12x}{4yz}\right)^{-3} =$

26) $\dfrac{8ab^{-7}}{-5c^{-3}} =$

27) $\left(-\dfrac{x^4}{x^5}\right)^{-5} =$

28) $\left(-\dfrac{x^{-2}}{7x^3}\right)^{-2} =$

29) $\left(-\dfrac{x^{-4}}{x^2}\right)^{-6} =$

# Scientific Notation

✎ **Write each number in scientific notation.**

1) 0.223 =

2) 0.09 =

3) 4.5 =

4) 900 =

5) 2,000 =

6) 0.006 =

7) 33 =

8) 9,400 =

9) 1,470 =

10) 52,000 =

11) 8,000,000 =

12) 0.00009 =

13) 2,158,000 =

14) 0.0039 =

15) 0.000075 =

16) 4,300,000 =

17) 130,000 =

18) 4,000,000,000 =

19) 0.00009 =

20) 0.0039 =

✎ **Write each number in standard notation.**

21) $4 \times 10^{-1} =$

22) $1.2 \times 10^{-3} =$

23) $2.7 \times 10^{5} =$

24) $6 \times 10^{-4} =$

25) $3.6 \times 10^{-3} =$

26) $5.5 \times 10^{5} =$

27) $3.2 \times 10^{4} =$

28) $3.88 \times 10^{6} =$

29) $7 \times 10^{-6} =$

30) $4.2 \times 10^{-7} =$

## Square Roots

✍ **Find the value each square root.**

1) $\sqrt{16}$ = ____

2) $\sqrt{25}$ = ____

3) $\sqrt{1}$ = ____

4) $\sqrt{64}$ = ____

5) $\sqrt{0}$ = ____

6) $\sqrt{196}$ = ____

7) $\sqrt{4}$ = ____

8) $\sqrt{256}$ = ____

9) $\sqrt{36}$ = ____

10) $\sqrt{289}$ = ____

11) $\sqrt{169}$ = ____

12) $\sqrt{144}$ = ____

13) $\sqrt{100}$ = ____

14) $\sqrt{1,600}$ = ____

15) $\sqrt{2,500}$ = ____

16) $\sqrt{324}$ = ____

17) $\sqrt{529}$ = ____

18) $\sqrt{20}$ = ____

19) $\sqrt{625}$ = ____

20) $\sqrt{18}$ = ____

21) $\sqrt{50}$ = ____

22) $\sqrt{1,024}$ = ____

23) $\sqrt{160}$ = ____

24) $\sqrt{32}$ = ____

✍ **Evaluate.**

25) $\sqrt{4} \times \sqrt{25}$ = _____

26) $\sqrt{36} \times \sqrt{49}$ = _____

27) $\sqrt{6} \times \sqrt{6}$ = _____

28) $\sqrt{13} \times \sqrt{13}$ = _____

29) $2\sqrt{5} \times 3\sqrt{5}$ = _____

30) $\sqrt{12} \times \sqrt{3}$ = _____

31) $\sqrt{13} + \sqrt{13}$ = _____

32) $\sqrt{10} + 2\sqrt{10}$ = _____

33) $12\sqrt{7} - 10\sqrt{7}$ = _____

34) $4\sqrt{10} \times 2\sqrt{10}$ = _____

35) $5\sqrt{3} \times 8\sqrt{3}$ = _____

36) $6\sqrt{3} - \sqrt{12}$ = _____

# Simplifying Radical Expressions

✎ **Simplify.**

1) $\sqrt{13x^2} =$

2) $\sqrt{75x^2} =$

3) $\sqrt[3]{27a} =$

4) $\sqrt{64x^5} =$

5) $\sqrt{216a} =$

6) $\sqrt[3]{63w^3} =$

7) $\sqrt{192x} =$

8) $\sqrt{125v} =$

9) $\sqrt[3]{128x^2} =$

10) $\sqrt{100x^9} =$

11) $\sqrt{16x^4} =$

12) $\sqrt[3]{500a^5} =$

13) $\sqrt{242} =$

14) $\sqrt{392p^3} =$

15) $\sqrt{8m^6} =$

16) $\sqrt{198x^3y^3} =$

17) $\sqrt{121x^5y^5} =$

18) $\sqrt{16a^6b^3} =$

19) $\sqrt{90x^5y^7} =$

20) $\sqrt[3]{64y^2x^6} =$

21) $10\sqrt{16x^4} =$

22) $6\sqrt{81x^2} =$

23) $\sqrt[3]{56x^2y^6} =$

24) $\sqrt[3]{1,000x^5y^7} =$

25) $8\sqrt{50a} =$

26) $\sqrt[4]{625x^8y} =$

27) $\sqrt{24x^4y^5r^3} =$

28) $5\sqrt{36x^4y^5z^8} =$

29) $3\sqrt[3]{343x^9y^7} =$

30) $5\sqrt{81a^5b^2c^9} =$

31) $\sqrt[4]{625x^8y^{16}} =$

# Multiplying Radical Expressions

✎ **Simplify.**

1) $\sqrt{5} \times \sqrt{5} =$

2) $\sqrt{5} \times \sqrt{10} =$

3) $\sqrt{3} \times \sqrt{12} =$

4) $\sqrt{49} \times \sqrt{47} =$

5) $\sqrt{7} \times -2\sqrt{28} =$

6) $3\sqrt{15} \times \sqrt{5} =$

7) $4\sqrt{72} \times \sqrt{2} =$

8) $\sqrt{5} \times -\sqrt{49} =$

9) $\sqrt{55} \times \sqrt{11} =$

10) $7\sqrt{42} \times 2\sqrt{216} =$

11) $\sqrt{45}(5 + \sqrt{5}) =$

12) $\sqrt{13x^2} \times \sqrt{13x^3} =$

13) $-2\sqrt{27} \times \sqrt{3} =$

14) $2\sqrt{13x^4} \times \sqrt{13x^4} =$

15) $\sqrt{14x^3} \times \sqrt{7x^2} =$

16) $-8\sqrt{5x} \times \sqrt{7x^5} =$

17) $-2\sqrt{16x^5} \times 4\sqrt{8x^3} =$

18) $-4\sqrt{32}(8 + \sqrt{32}) =$

19) $\sqrt{32x}(10 - \sqrt{2x}) =$

20) $\sqrt{2x}(8\sqrt{x^5} + \sqrt{8}) =$

21) $\sqrt{20r}(5 + \sqrt{5}) =$

22) $-4\sqrt{7x} \times 3\sqrt{14x^5} =$

23) $-2\sqrt{12x} \times 3\sqrt{2x}$

24) $-\sqrt{7v^3}(-3\sqrt{42v}) =$

25) $(\sqrt{11} - 5)(\sqrt{11} + 5) =$

26) $(-3\sqrt{5} + 3)(\sqrt{5} - 4) =$

27) $(4 - 6\sqrt{3})(-6 + \sqrt{3}) =$

28) $(8 - 3\sqrt{5})(7 - \sqrt{5}) =$

29) $(-1 - \sqrt{3x})(4 + \sqrt{3x}) =$

30) $(-5 + 2\sqrt{7r})(-5 + \sqrt{7r}) =$

31) $(-\sqrt{7n} + 1)(-\sqrt{7} - 5) =$

32) $(-3 + \sqrt{3})(5 - 2\sqrt{3x}) =$

# Simplifying Radical Expressions Involving Fractions

✎ **Simplify.**

1) $\dfrac{\sqrt{5}}{\sqrt{3}} =$

2) $\dfrac{\sqrt{18}}{\sqrt{45}} =$

3) $\dfrac{\sqrt{10}}{5\sqrt{2}} =$

4) $\dfrac{13}{\sqrt{3}} =$

5) $\dfrac{12\sqrt{5r}}{\sqrt{m^5}} =$

6) $\dfrac{11\sqrt{2}}{\sqrt{k}} =$

7) $\dfrac{6\sqrt{20x^3}}{\sqrt{16x}} =$

8) $\dfrac{\sqrt{14x^3y^4}}{\sqrt{7x^4y^3}} =$

9) $\dfrac{1}{1-\sqrt{5}} =$

10) $\dfrac{1-8\sqrt{a}}{\sqrt{11a}} =$

11) $\dfrac{\sqrt{a}}{\sqrt{a}+\sqrt{b}} =$

12) $\dfrac{1-\sqrt{5}}{2-\sqrt{6}} =$

13) $\dfrac{4+\sqrt{7}}{3-\sqrt{8}} =$

14) $\dfrac{5}{-3-3\sqrt{3}} =$

15) $\dfrac{7}{2-\sqrt{5}} =$

16) $\dfrac{\sqrt{7}-\sqrt{3}}{\sqrt{3}-\sqrt{7}} =$

17) $\dfrac{\sqrt{5}+\sqrt{7}}{\sqrt{7}-\sqrt{5}} =$

18) $\dfrac{2\sqrt{2}-\sqrt{3}}{3\sqrt{2}+\sqrt{5}} =$

19) $\dfrac{\sqrt{11}+5\sqrt{3}}{4-\sqrt{11}} =$

20) $\dfrac{\sqrt{5}+\sqrt{3}}{2-\sqrt{3}} =$

21) $\dfrac{\sqrt{32a^7b^4}}{\sqrt{2ab^3}} =$

22) $\dfrac{10\sqrt{21x^5}}{5\sqrt{x^3}} =$

## Adding and Subtracting Radical Expressions

✎ **Simplify.**

1) $\sqrt{2} + \sqrt{8} =$

2) $3\sqrt{50} + 4\sqrt{2} =$

3) $2\sqrt{12} - 4\sqrt{3} =$

4) $5\sqrt{32} - 5\sqrt{2} =$

5) $3\sqrt{75} - 5\sqrt{3} =$

6) $-\sqrt{72} - 4\sqrt{2} =$

7) $-7\sqrt{16} - 4\sqrt{25} =$

8) $8\sqrt{24} + 2\sqrt{6} =$

9) $10\sqrt{49} - 7\sqrt{100} =$

10) $-7\sqrt{5} + 9\sqrt{45} =$

11) $-15\sqrt{12} + 14\sqrt{48} =$

12) $20\sqrt{4} - 2\sqrt{25} =$

13) $-2\sqrt{20} + 7\sqrt{5} =$

14) $8\sqrt{7} - 2\sqrt{63} =$

15) $5\sqrt{44} + 3\sqrt{11} =$

16) $3\sqrt{27} - 5\sqrt{48} =$

17) $\sqrt{144} - \sqrt{81} =$

18) $3\sqrt{20} - 6\sqrt{5} =$

19) $-2\sqrt{7} + 8\sqrt{28} =$

20) $3\sqrt{75} - 2\sqrt{3} =$

21) $5\sqrt{27} - 3\sqrt{3} =$

22) $-7\sqrt{30} + 6\sqrt{120} =$

23) $-7\sqrt{24} - 2\sqrt{6} =$

24) $-\sqrt{32x} + 4\sqrt{2x} =$

25) $\sqrt{7y^2} + y\sqrt{112} =$

26) $\sqrt{45mn^2} + 2n\sqrt{5m} =$

27) $-4\sqrt{12a} - 4\sqrt{3a} =$

28) $-5\sqrt{15ab} - 2\sqrt{60ab} =$

29) $\sqrt{45x^2y} + x\sqrt{20y} =$

30) $2\sqrt{7a} + 4\sqrt{63a} =$

# Answers of Worksheets

## Multiplication Property of Exponents

1) $4^6$
2) $8^5$
3) $7^6$
4) $9^4$
5) $2^7$
6) $5^7$
7) $4^7$
8) $5x^2$

9) $x^6$
10) $x^9$
11) $x^9$
12) $30x^2$
13) $16x^6$
14) $7x^4$
15) $12x^6$
16) $5x^8$

17) $4x^9$
18) $6x^6$
19) $24x^8$
20) $20x^4y$
21) $7x^{10}y^5$
22) $x^7y^7$
23) $12x^8y^4$
24) $36x^6y^5$

25) $30x^{11}y^6$
26) $32x^4y^9$
27) $12x^4y^8$
28) $16x^7y^{10}$
29) $21x^3y^{13}$
30) $4x^7y^4$
31) $27x^6y^8$
32) $120x^9y^6$

## Zero and Negative Exponents

1) $1$
2) $\frac{1}{4}$
3) $0$
4) $1$
5) $\frac{1}{25}$
6) $\frac{1}{27}$
7) $\frac{1}{9}$
8) $\frac{1}{100}$
9) $\frac{1}{144}$
10) $\frac{1}{32}$
11) $\frac{1}{81}$

12) $\frac{1}{16}$
13) $\frac{1}{216}$
14) $\frac{1}{1,000}$
15) $\frac{1}{30}$
16) $\frac{1}{225}$
17) $\frac{1}{64}$
18) $\frac{1}{128}$
19) $\frac{1}{125}$
20) $\frac{1}{256}$
21) $\frac{1}{243}$

22) $\frac{1}{10,000}$
23) $\frac{1}{1,024}$
24) $\frac{1}{512}$
25) $\frac{1}{400}$
26) $\frac{1}{196}$
27) $\frac{1}{729}$
28) $\frac{1}{10,000}$
29) $\frac{1}{625}$
30) $\frac{1}{4,096}$
31) $64$
32) $36$

33) $49$
34) $\frac{27}{8}$
35) $169$
36) $\frac{144}{49}$
37) $216$
38) $90,000$
39) $\frac{81}{4}$
40) $\frac{5}{7}$
41) $1$
42) $1,024$

## Division Property of Exponents

1) $\frac{1}{5}$
2) $8^2$
3) $4^4$
4) $\frac{1}{3^4}$

5) $\frac{1}{x^5}$
6) $\frac{1}{3^4}$
7) $9^2$
8) $10$

9) $1$
10) $\frac{1}{2x^5}$
11) $\frac{3x^5}{4}$
12) $\frac{3}{2x}$

13) $\frac{7x^5}{y^9}$
14) $\frac{9y^3}{x^4}$
15) $\frac{2x^6}{9}$

16) $\frac{7y^6}{x}$  18) $\frac{5}{x^2}$  19) $\frac{1}{2x^5y^3}$  21) $\frac{9}{4x^6}$

17) $\frac{2}{x^4y^{12}}$  20) $\frac{1}{7}$

## Powers of Products and Quotients

1) $4^6$

2) $2^{12}$

3) $2^8$

4) $5^{36}$

5) $19^{18}$

6) $2^{28}$

7) $5^6$

8) $4^{16}$

9) $64x^{10}$

10) $81x^8y^{16}$

11) $49x^{10}y^4$

12) $125x^{12}y^{12}$

13) $32x^{15}y^{15}$

14) $1,000x^9y^{12}$

15) $169y^8$

16) $25x^{20}$

17) $216x^{21}y^{18}$

18) $144x^{24}$

19) $256x^{20}$

20) $32x^{20}y^{15}$

21) $225x^{14}y^4$

22) $512x^9y^{15}$

23) $1,296x^4y^8$

24) $\frac{16}{x^8}$

25) $x^9$

26) $\frac{216y^3}{x^{12}}$

27) $\frac{1}{x^6y^{12}}$

28) $x^6y^6$

29) $\frac{25y^{16}}{x^4}$

30) $\frac{16}{y^{12}}$

## Negative Exponents and Negative Bases

1) $-\frac{1}{9}$

2) $-\frac{1}{81}$

3) $-\frac{1}{32}$

4) $-\frac{1}{x^7}$

5) $\frac{11}{x}$

6) $-\frac{8}{x^3}$

7) $-\frac{12}{x^5}$

8) $-\frac{9}{x^8y^6}$

9) $\frac{32}{x^5y}$

10) $\frac{10}{a^9b^3}$

11) $-\frac{17x^4}{y^6}$

12) $-25x^5$

13) $-13xa^7$

14) $81$

15) $\frac{16}{9}$

16) $-14a^6b^3$

17) $-7x^9$

18) $-\frac{b^5}{a^9}$

19) $-11x^5$

20) $-\frac{bc^6}{2}$

21) $12a^5b^4$

22) $-\frac{p^7}{4n^4}$

23) $-\frac{8ac^5}{3b^6}$

24) $\frac{c^4}{16a^4}$

25) $\frac{y^3z^3}{27x^3}$

26) $-\frac{8ac^3}{5b^7}$

27) $-x^5$

28) $49x^{10}$

29) $x^{36}$

## Scientific Notation

1) $2.23 \times 10^{-1}$

2) $9 \times 10^{-2}$

3) $4.5 \times 10^{0}$

4) $9 \times 10^{2}$

5) $2 \times 10^{3}$

6) $6 \times 10^{-3}$

7) $3.3 \times 10^{1}$

8) $9.4 \times 10^{3}$

9) $1.47 \times 10^{3}$

10) $5.2 \times 10^{4}$

11) $8 \times 10^{6}$

12) $9 \times 10^{-5}$

13) $2.158 \times 10^{6}$

14) $3.9 \times 10^{-3}$

15) $7.5 \times 10^{-5}$

16) $4.3 \times 10^{6}$

17) $1.3 \times 10^{5}$

18) $4 \times 10^{9}$

19) $9 \times 10^{-5}$

20) $3.9 \times 10^{-3}$

21) 0.4

22) 0.0012

23) 270,000

24) 0.0006

25) 0.0036

26) 550,000

27) 32,000

28) 3,880,000

29) 0.000007

30) 0.00000042

**Square Roots**

1) 4

2) 5

3) 1

4) 8

5) 0

6) 14

7) 2

8) 16

9) 6

10) 17

11) 13

12) 12

13) 10

14) 40

15) 50

16) 18

17) 23

18) $2\sqrt{5}$

19) 25

20) $3\sqrt{2}$

21) $5\sqrt{2}$

22) 32

23) $4\sqrt{10}$

24) $4\sqrt{2}$

25) 10

26) 42

27) 6

28) 13

29) 30

30) 6

31) $2\sqrt{13}$

32) $3\sqrt{10}$

33) $2\sqrt{7}$

34) 80

35) 120

36) $4\sqrt{3}$

**Simplifying radical expressions**

1) $x\sqrt{13}$

2) $5x\sqrt{3}$

3) $3\sqrt[3]{a}$

4) $8x^2\sqrt{x}$

5) $6\sqrt{6a}$

6) $w\sqrt[3]{63}$

7) $8\sqrt{3x}$

8) $5\sqrt{5v}$

9) $4\sqrt[3]{2x^2}$

10) $10x^4\sqrt{x}$

11) $4x^2$

12) $5a\sqrt[3]{4a^2}$

13) $11\sqrt{2}$

14) $14p\sqrt{2p}$

15) $2m^3\sqrt{2}$

16) $3x \cdot y\sqrt{22xy}$

17) $11x^2y^2\sqrt{xy}$

18) $4a^3b\sqrt{b}$

19) $3x^2y^3\sqrt{10xy}$

20) $4x^2\sqrt[3]{y^2}$

21) $40x^2$

22) $54x$

23) $2y^2\sqrt[3]{7x^2}$

24) $10xy^2\sqrt[3]{x^2y}$

25) $40\sqrt{2a}$

26) $5x^2\sqrt[4]{y}$

27) $2x^2y^2r\sqrt{6yr}$

28) $30x^2y^2z^4\sqrt{y}$

29) $21x^3y^2\sqrt[3]{y}$

30) $45a^2bc^4\sqrt{ac}$

31) $5x^2y^4$

## Multiplying radical expressions

1) $5$

2) $5\sqrt{2}$

3) $6$

4) $7\sqrt{47}$

5) $-28$

6) $15\sqrt{3}$

7) $48$

8) $-5\sqrt{7}$

9) $11\sqrt{5}$

10) $504\sqrt{7}$

11) $15\sqrt{5} + 15$

12) $13x^2\sqrt{x}$

13) $-18$

14) $26x^4$

15) $7x^2\sqrt{2x}$

16) $-8x^3\sqrt{35}$

17) $-64x^4\sqrt{2}$

18) $-128\sqrt{2} - 128$

19) $40\sqrt{2x} - 8x$

20) $8x^3\sqrt{2} + 4\sqrt{x}$

21) $10\sqrt{5r} + 10\sqrt{r}$

22) $-84x^3\sqrt{2}$

23) $-12\sqrt{6}x$

24) $21v^2\sqrt{6}$

25) $-14$

26) $15\sqrt{5} - 27$

27) $40\sqrt{3} - 42$

28) $71 - 29\sqrt{5}$

29) $-3x - 5\sqrt{3x} - 4$

30) $14r - 15\sqrt{7r} + 25$

31) $7\sqrt{n} + 5\sqrt{7n} - \sqrt{7} - 5$

32) $-15 + 6\sqrt{3x} + 5\sqrt{3} - 6\sqrt{x}$

## Simplifying radical expressions involving fractions

1) $\frac{\sqrt{15}}{3}$

2) $\frac{9\sqrt{10}}{45} = \frac{\sqrt{10}}{5}$

3) $\frac{\sqrt{20}}{10} = \frac{\sqrt{5}}{5}$

4) $\frac{13\sqrt{3}}{3}$

5) $\frac{12\sqrt{5mr}}{m^3}$

6) $\frac{11\sqrt{2k}}{k}$

7) $3x\sqrt{5}$

8) $\frac{\sqrt{2x}}{xy}$

9) $\frac{-1-\sqrt{5}}{4}$

10) $\frac{\sqrt{11a} - 8a\sqrt{11}}{11a}$

11) $\frac{a-\sqrt{ab}}{a-b}$

12) $\frac{\sqrt{30}+2\sqrt{5}-\sqrt{6}-2}{2}$

13) $12 + 8\sqrt{2} + 3\sqrt{7} + 2\sqrt{14}$

14) $-\frac{5(\sqrt{3}-1)}{6}$

15) $-14 - 7\sqrt{5}$

16) $-1$

17) $6 + \sqrt{35}$

18) $\frac{12 - 2\sqrt{10} - 3\sqrt{6} + \sqrt{15}}{13}$

19) $\frac{4\sqrt{11}+11+20\sqrt{3}+5\sqrt{33}}{5}$

20) $2\sqrt{5} + 3 + \sqrt{15} + 2\sqrt{3}$

21) $4a^3\sqrt{b}$

22) $2x\sqrt{21}$

## Adding and subtracting radical expressions

1) $3\sqrt{2}$

2) $19\sqrt{2}$

3) $0$

4) $15\sqrt{2}$

5) $10\sqrt{3}$

6) $-10\sqrt{2}$

7) $-48$

8) $18\sqrt{6}$

9) $0$

10) $20\sqrt{5}$

11) $26\sqrt{3}$

12) $30$

13) $3\sqrt{5}$

14) $2\sqrt{7}$

15) $13\sqrt{11}$

16) $-11\sqrt{3}$

17) $3$

18) $0$

19) $14\sqrt{7}$

20) $13\sqrt{3}$

21) $12\sqrt{3}$

22) $5\sqrt{30}$

23) $-16\sqrt{6}$

24) $0$

25) $5y\sqrt{7}$

26) $5n\sqrt{5m}$

27) $-12\sqrt{3a}$

28) $-9\sqrt{15ab}$

29) $5x\sqrt{5y}$

30) $14\sqrt{7a}$

# Chapter 5 :

# Algebraic Expressions

## Topics that you'll practice in this chapter:

- ✓ Simplifying Variable Expressions
- ✓ Simplifying Polynomial Expressions
- ✓ Translate Phrases into an Algebraic Statement
- ✓ The Distributive Property
- ✓ Evaluating One Variable Expressions
- ✓ Evaluating Two Variables Expressions
- ✓ Combining like Terms

*I want freedom for the full expression of my personality.*
*Mahatma Gandhi*

# Simplifying Variable Expressions

✎ **Simplify each expression.**

1) $3(x + 5) =$

2) $(-4)(7x - 5) =$

3) $11x + 5 - 6x =$

4) $-4 - 2x^2 - 6x^2 =$

5) $7 + 13x^2 + 3 =$

6) $3x^2 + 7x + 15x^2 =$

7) $3x^2 - 12x^2 + 4x =$

8) $4x^2 - 8x - 2x =$

9) $6x + 7(3 - 4x) =$

10) $8x + 4(15x - 3) =$

11) $6(-3x - 9) - 17 =$

12) $-11x^2 - (-5x) =$

13) $2x + 7 + 5 - 8x =$

14) $7 + 6x - 11 - 5x =$

15) $27x + 8 - 13 - 5x =$

16) $(-11)(-5x + 2) - 41x =$

17) $19x - 4(4 - 2x) =$

18) $16x + 3(3x + 6) + 10 =$

19) $5(-2x - 4) - 13x =$

20) $16x - 3x(x + 10) =$

21) $17x + 5x(2 - 4x) =$

22) $5x(-4x - 7) + 20x =$

23) $25x - 19 + 4x^2 =$

24) $6x(x - 11) + 25 =$

25) $4x - 5 + 15x + 3x^2 =$

26) $-7x^2 - 11x - 9x =$

27) $10x - 9x^2 - 3x^2 - 7 =$

28) $13 + 3x^2 - 9x^2 - 21x =$

29) $22x + 10x^2 - 15x + 17 =$

30) $4x^2 + 25x + 21x^2 =$

31) $29 - 12x^2 - 23x - 4x^2 =$

32) $22x - 19x - 9x^2 + 30 =$

## Simplifying Polynomial Expressions

✎ Simplify each polynomial.

1) $(2x^3 + 8x^2) - (11x + 3x^2) =$ _____

2) $(2x^5 + 7x^3) - (5x^3 + 11x^2) =$ _____

3) $(41x^4 + 5x^2) - (4x^2 + 20x^4) =$ _____

4) $13x - 8x^2 + 4(4x^2 + 3x^3) =$ _____

5) $(4x^3 - 22) + 5(3x^2 - 6x^3) =$ _____

6) $(4x^3 - 3x) - 5(2x^3 + x^4) =$ _____

7) $5(5x - 2x^3) - 2(8x^3 + 5x^2) =$ _____

8) $(3x^2 - 10x) - (5x^3 + 14x^2) =$ _____

9) $5x^3 - (3x^4 + 5x) + 2x^2 =$ _____

10) $11x^4 - (3x^2 + 5x) + 7x =$ _____

11) $(6x^2 - 3x^4) - (10x^4 + 3x^2) =$ _____

12) $2x^2 - 7x^3 + 19x^4 - 22x^3 =$ _____

13) $10x^2 - x^4 + 4x^4 - 32x^3 =$ _____

14) $-5x^2 + 17x^3 - 8x^2 - 6x =$ _____

15) $x^4 - 11x^5 - 30x^4 + 5x^2 =$ _____

16) $21x^3 + 13x - 5x^2 - 11x^3 =$ _____

# Translate Phrases into an Algebraic Statement

✎ **Write an algebraic expression for each phrase.**

1) 9 multiplied by $x$. _____

2) Subtract 11 from $y$. _____

3) 19 divided by $x$. _____

4) 38 decreased by $y$. _____

5) Add $y$ to 40. _____

6) The square of 6. _____

7) $x$ raised to the fifth power. _____

8) The sum of six and a number. _____

9) The difference between fifty–seven and $y$. _____

10) The quotient of nine and a number. _____

11) The quotient of the square of $x$ and 25. _____

12) The difference between $x$ and 6 is 19. _____

13) 10 times $a$ reduced by the square of $b$. _____

14) Subtract the product of $a$ and $b$ from 41. _____

# The Distributive Property

✎ Use the distributive property to simply each expression.

1) $4(1 + 2x) =$

2) $2(4 + 7x) =$

3) $3(4x - 4) =$

4) $(2x - 5)(-6) =$

5) $(-3)(x + 6) =$

6) $(4 + 3x)2 =$

7) $(-5)(8 - 3x) =$

8) $-(-5 - 7x) =$

9) $(-6x + 3)(-3) =$

10) $(-4)(x - 7) =$

11) $-(5 - 3x) =$

12) $3(9 + 4x) =$

13) $6(4 + 3x) =$

14) $(-5x + 3)2 =$

15) $(5 - 8x)(-3) =$

16) $(-12)(3x + 3) =$

17) $(5 - 3x)6 =$

18) $4(2 + 6x) =$

19) $8(7x - 3) =$

20) $(-2x + 3)4 =$

21) $(7 - 5x)(-9) =$

22) $(-10)(x - 8) =$

23) $(11 - 4x)3 =$

24) $(-6)(10x - 4) =$

25) $(3 - 9x)(-7) =$

26) $(-9)(x + 9) =$

27) $(-3 + 5x)(-7) =$

28) $(-5)(8 - 10x) =$

29) $12(4x - 8) =$

30) $(-10x + 13)(-3) =$

31) $(-8)(3x - 2) + 4(x + 5) =$

32) $(-8)(x + 4) - (6 + 5x) =$

## Evaluating One Variable Expressions

✎ **Evaluate each expression using the value given.**

1) $8 - x$, $x = 5$

2) $x - 9$, $x = 5$

3) $5x + 4$, $x = 3$

4) $x - 13$, $x = -4$

5) $12 - x$, $x = 4$

6) $x + 2$, $x = 6$

7) $4x + 8$, $x = 3$

8) $x + (-7)$, $x = -8$

9) $4x + 5$, $x = 2$

10) $3x + 9$, $x = -2$

11) $15 + 3x - 7$, $x = 2$

12) $17 - 3x$, $x = 3$

13) $8x - 9$, $x = 4$

14) $5x + 4$, $x = -3$

15) $10x + 5$, $x = 3$

16) $14 - 4x$, $x = -6$

17) $3(5x + 3)$, $x = 9$

18) $4(-3x - 6)$, $x = 3$

19) $7x - 2x + 12$, $x = 4$

20) $(5x + 6) \div 2$, $x = 8$

21) $(x + 18) \div 10$, $x = 12$

22) $5x - 12 + 3x$, $x = -3$

23) $(6 - 4x)(-3)$, $x = -4$

24) $9x^2 + 3x - 6$, $x = 2$

25) $x^2 - 10x$, $x = -5$

26) $3x(7 - 2x)$, $x = 2$

27) $12x + 6 - 2x^2$, $x = -4$

28) $(-3)(4x - 8 + 3x)$, $x = 3$

29) $(-6) + \frac{x}{4} + 3x$, $x = 16$

30) $(-6) + \frac{x}{5}$, $x = 35$

31) $\left(-\frac{45}{x}\right) - 7 + 2x$, $x = 9$

32) $\left(-\frac{21}{x}\right) - 12 + 4x$, $x = 7$

# Evaluating Two Variables Expressions

✎ **Evaluate each expression using the values given.**

1) $2x - 4y$,

   $x = 4, y = 1$

2) $3x + 5y$,

   $x = -2, y = 2$

3) $-7a + 4b$,

   $a = 2, b = 4$

4) $3x + 5 - y$,

   $x = 5, y = 6$

5) $3z + 12 - 2k$,

   $z = 5, k = 6$

6) $6(-x - 3y)$,

   $x = 5, y = -2$

7) $5a + 3b$,

   $a = 3, b = 4$

8) $7x \div 3y$,

   $x = 3, y = 7$

9) $2x + 15 + 5y$,

   $x = -3, y = 1$

10) $5a - (18 - b)$,

    $a = 2, b = 8$

11) $2z + 20 + 5k$,

    $z = -6, k = 5$

12) $xy + 10 + 4x$,

    $x = 3, y = 5$

13) $2x + 4y - 8 + 5$,

    $x = 5, y = 2$

14) $\left(-\frac{24}{x}\right) + 3 + 2y$,

    $x = 4, y = 6$

15) $(-3)(-3a - 3b)$,

    $a = 4, b = 5$

16) $12 + 4x - 7 - y$,

    $x = 3, y = 5$

17) $11x + 5 - 8y + 6$,

    $x = 5, y = 2$

18) $10 + 2(-4x - 5y)$,

    $x = 5, y = 4$

19) $5x + 13 + 6y$,

    $x = 5, y = 6$

20) $10a - (7a + 3b) - 11$,

    $a = 3, b = 8$

## Combining like Terms

✍ **Simplify each expression.**

1) $11x + 3x + 6 =$

2) $8(2x - 6) =$

3) $18x - 7x + 11 =$

4) $(-4)(6x - 7) =$

5) $22x - 10x - 5 =$

6) $32x - 13 + 8x =$

7) $15 - (8x - 11) =$

8) $-24x + 17 - 11x =$

9) $12x - 8 - 6x + 9 =$

10) $21x + 5 - 36 + 12x =$

11) $28x + 3x - 11 =$

12) $(-3x + 4)5 =$

13) $2 + 4x + 9x - 8 =$

14) $6(2x - 5x) - 4 =$

15) $4(5x + 11) + 3x =$

16) $x - 14 - 11x =$

17) $5(10 + 9x) - 8x =$

18) $42x + 17 - 23x =$

19) $(-7x) + 19 + 20x =$

20) $(-7x) - 33 + 29x =$

21) $4(5x + 3) - 19x =$

22) $5(6 - 2x) - 15x =$

23) $-24x + (11 - 18x) =$

24) $(-9) - (6)(7x + 3) =$

25) $(-1)(8x - 10) - 21x =$

26) $-36x + 14 + 27x - 5x =$

27) $3(-13x + 6) - 17x =$

28) $-5x - 42 + 32x =$

29) $37x - 19x + 15 - 9x =$

30) $3(5x + 7x) - 31 =$

31) $14 - 6x - 15 - 9x =$

32) $-2(-5x - 7x) + 27x =$

# Answers of Worksheets

**Simplifying Variable Expressions**

1) $3x + 15$

2) $-28x + 20$

3) $5x + 5$

4) $-8x^2 - 4$

5) $13x^2 + 10$

6) $18x^2 + 7x$

7) $-9x^2 + 4x$

8) $4x^2 - 10x$

9) $-22x + 21$

10) $68x - 12$

11) $-18x - 71$

12) $-11x^2 + 5x$

13) $-6x + 12$

14) $x - 4$

15) $22x - 5$

16) $14x - 22$

17) $27x - 16$

18) $25x + 28$

19) $-23x - 20$

20) $-3x^2 - 14x$

21) $-20x^2 + 27x$

22) $-20x^2 - 15x$

23) $4x^2 + 25x - 19$

24) $6x^2 - 66x + 25$

25) $3x^2 + 19x - 5$

26) $-7x^2 - 20x$

27) $-12x^2 + 10x - 7$

28) $-6x^2 - 21x + 13$

29) $10x^2 + 7x + 17$

30) $25x^2 + 25x$

31) $-16x^2 - 23x + 29$

32) $-9x^2 + 3x + 30$

**Simplifying Polynomial Expressions**

1) $2x^3 + 5x^2 - 11x$

2) $2x^5 + 2x^3 - 11x^2$

3) $21x^4 + x^2$

4) $12x^3 + 8x^2 + 13x$

5) $-26x^3 + 15x^2 - 22$

6) $-5x^4 - 6x^3 - 3x$

7) $-26x^3 - 10x^2 + 25x$

8) $-5x^3 - 11x^2 - 10x$

9) $-3x^4 + 5x^3 + 2x^2 - 5x$

10) $11x^4 - 3x^2 + 2x$

11) $-13x^4 + 3x^2$

12) $19x^4 - 29x^3 + 2x^2$

13) $3x^4 - 32x^3 + 10x^2$

14) $17x^3 - 13x^2 - 6x$

15) $-11x^5 - 29x^4 + 5x^2$

16) $10x^3 - 5x^2 + 13x$

**Translate Phrases into an Algebraic Statement**

1) $9x$

2) $y - 11$

3) $\frac{19}{x}$

4) $38 - y$

5) $y + 40$

6) $6^2$

7) $x^5$

8) $6 + x$

9) $57 - y$

10) $\frac{9}{x}$

11) $\frac{x^2}{25}$

12) $x - 6 = 19$

13) $10a - b^2$

14) $41 - ab$

**The Distributive Property**

1) $8x + 4$

2) $14x + 8$

3) $12x - 12$

4) $-12x + 30$

5) $-3x - 18$

6) $6x + 8$

7) $15x - 40$

8) $7x + 5$

9) $18x - 9$

10) $-4x + 28$

11) $3x - 5$

12) $12x + 27$

13) $18x + 24$

14) $-10x + 6$

15) $24x - 15$

16) $-36x - 36$

17) $-18x + 30$

18) $24x + 8$

19) $56x - 24$

20) $-8x + 12$

21) $45x - 63$

22) $-10x + 80$

23) $-12x + 33$

24) $-60x + 24$

25) $63x - 21$

26) $-9x - 81$

27) $-35x + 21$

28) $50x - 40$

29) $48x - 96$

30) $30x - 39$

31) $-20x + 36$

32) $-13x - 38$

**Evaluating One Variables**

1) 3

2) −4

3) 19

4) −17

5) 8

6) 8

7) 20

8) − 15

9) 13

10) 3

11) 14

12) 8

13) 23

14) −11

15) 35

16) 38

17) 144

18) −60

19) 32

20) 23

21) 3

22) −36

23) −66

24) 36

25) 75

26) 18

27) −74

28) −39

29) 46

30) 1

31) 6

32) 13

**Evaluating Two Variables**

1) 4

2) 4

3) 2

4) 14

5) 15

6) 6

7) 27

8) 1

9) 14

10) 0

11) 33

12) 37

13) 15

14) 9

15) 81

16) 12

17) 50

18) −70

19) 74

20) −26

**Combining like Terms**

1) $14x + 6$

2) $16x - 48$

3) $11x + 11$

4) $-24x + 28$

5) $12x - 5$

6) $40x - 13$

7) $-8x + 26$

8) $-35x + 17$

9) $6x + 1$

10) $33x - 31$

11) $31x - 11$

12) $-15x + 20$

13) $13x - 6$

14) $-18x - 4$

15) $23x + 44$

16) $-10x - 14$

17) $37x + 50$

18) $19x + 17$

19) $13x + 19$

20) $22x - 33$

21) $x + 12$

22) $-25x + 30$

23) $-42x + 11$

24) $-42x - 27$

25) $-29x + 10$

26) $-14x + 14$

27) $-56x + 18$

28) $27x - 42$

29) $9x + 15$

30) $36x - 31$

31) $-15x - 1$

32) $51x$

# Chapter 6 :

# Equations and Inequalities

## Topics that you'll practice in this chapter:

- ✓ One–Step Equations
- ✓ Multi–Step Equations
- ✓ Graphing Single–Variable Inequalities
- ✓ One–Step Inequalities
- ✓ Multi-Step Inequalities
- ✓ Systems of Equations
- ✓ Systems of Equations Word Problems

*"Life is a math equation. In order to gain the most, you have to know how to convert negatives into positives." – Anonymous*

## One–Step Equations

✍ **Find the answer for each equation.**

1) $3x = 90, x =$ ____

2) $5x = 35, x =$ ____

3) $6x = 24, x =$ ____

4) $24x = 144, x =$ ____

5) $x + 15 = 20, x =$ ____

6) $x - 7 = 4, x =$ ____

7) $x - 9 = 2, x =$ ____

8) $x + 15 = 23, x =$ ____

9) $x - 4 = 13, x =$ ____

10) $12 = 16 + x, x =$ ____

11) $x - 10 = 2, x =$ ____

12) $5 - x = -11, x =$ ____

13) $28 = -6 + x, x =$ ____

14) $x - 20 = -35, x =$ ____

15) $x + 14 = -4, x =$ ____

16) $14 = 28 - x, x =$ ____

17) $7 + x = -7, x =$ ____

18) $x - 16 = 4, x =$ ____

19) $30 = x - 15, x =$ ____

20) $x - 5 = -18, x =$ ____

21) $x - 10 = 24, x =$ ____

22) $x - 20 = -25, x =$ ____

23) $x - 17 = 30, x =$ ____

24) $-70 = x - 28, x =$ ____

25) $x - 9 = 13, x =$ ____

26) $36 = 4x, x =$ ____

27) $x - 35 = 25, x =$ ____

28) $x - 25 = 10, x =$ ____

29) $70 - x = 16, x =$ ____

30) $x - 10 = 14, x =$ ____

31) $17 - x = -13, x =$ __

32) $x - 9 = -30, x =$ ____

## Multi–Step Equations

✍ **Find the answer for each equation.**

1) $3x + 3 = 9$

2) $-x + 5 = 12$

3) $4x - 8 = 8$

4) $-(3 - x) = 5$

5) $4x - 8 = 16$

6) $12x - 15 = 9$

7) $2x - 18 = 2$

8) $4x + 8 = 16$

9) $24x + 27 = 75$

10) $-14(3 + x) = 14$

11) $-3(2 + x) = 6$

12) $12 = -(x - 7)$

13) $3(3 - x) = 30$

14) $-15 = -(3x + 6)$

15) $40(3 + x) = 40$

16) $5(x - 10) = 25$

17) $-18 = x + 8x$

18) $3x + 25 = -2x - 10$

19) $7(6 + 3x) = -63$

20) $18 - 3x = -4 - 5x$

21) $4 - 6x = 36 + 2x$

22) $15 + 15x = -5 + 5x$

23) $42 = (-6x) - 7 + 7$

24) $21 = 3x - 21 + 4x$

25) $-18 = -6x - 9 + 3x$

26) $5x - 15 = -29 + 6x$

27) $7x - 18 = 4x + 3$

28) $-7 - 4x = 5(4 - x)$

29) $x - 5 = -5(-3 - x)$

30) $13x - 68 = 15x - 102$

31) $-5x - 3 = -3(9 + 3x)$

32) $-2x - 15 = 6x + 17$

# Graphing Single–Variable Inequalities

✎ **Draw a graph for each inequality.**

1) $x > -1$

2) $x \leq 2$

3) $x \geq 0$

4) $x < -3$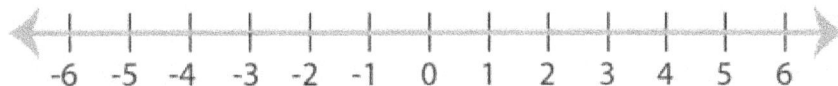

5) $x < \frac{1}{2}$

6) $x \leq -2$

7) $x \leq 3$

8) $x \geq -\frac{7}{2}$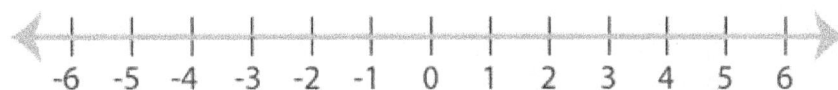

## One–Step Inequalities

✎ **Find the answer for each inequality and graph it.**

1) $x + 4 \geq 4$

2) $x - 5 \leq 2$

3) $5x > 35$

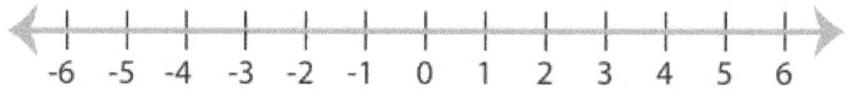

4) $9 + x \leq 11$

5) $x - 5 < -9$

6) $9x \geq 72$

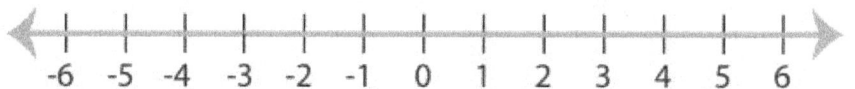

7) $9x \leq 27$

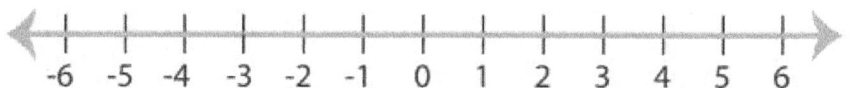

8) $x + 19 > 16$

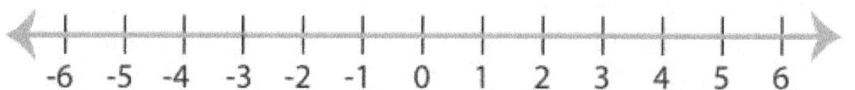

# Multi-Step Inequalities

✎ Calculate each inequality.

1) $x - 3 \leq 7$

2) $8 - x \leq 8$

3) $3x - 9 \leq 9$

4) $4x - 4 \geq 8$

5) $x - 7 \geq 1$

6) $5x - 15 \leq 5$

7) $6x - 8 \leq 4$

8) $-11 + 6x \leq 12$

9) $4(x - 4) \leq 16$

10) $3x - 10 \leq 11$

11) $5x - 25 < 25$

12) $9x - 5 < 22$

13) $20 - 7x \geq -15$

14) $33 + 6x < 45$

15) $8 + 8x \geq 96$

16) $7 + 3x < 13$

17) $4x - 3 < 9$

18) $5(2 - 2x) \geq -30$

19) $-(7 + 6x) < 29$

20) $12 - 8x \geq -20$

21) $-4(x - 6) > 24$

22) $\dfrac{3x + 9}{6} \leq 10$

23) $\dfrac{4x - 10}{3} \leq 2$

24) $\dfrac{2x - 8}{3} > 2$

25) $8 + \dfrac{x}{6} < 9$

26) $\dfrac{9x}{7} - 4 < 5$

27) $\dfrac{15x + 45}{15} > 1$

28) $16 + \dfrac{x}{4} < 6$

# Systems of Equations

✎ **Calculate each system of equations.**

1) $-x + y = 2$      $x = $ ___
   $-4x + 2y = 6$      $y = $ ___

2) $-15x + 3y = -9$      $x = $ ___
   $9x - 16y = 48$      $y = $ ___

3) $y = -7$      $x = $ ___
   $6x + 5y = 7$      $y = $ ___

4) $3y = -9x + 15$      $x = $ ___
   $5x - 4y = -3$      $y = $ ___

5) $10x - 9y = -13$      $x = $ ___
   $-5x + 3y = 11$      $y = $ ___

6) $-12x - 16y = 20$      $x = $ ___
   $6x - 12y = 30$      $y = $ ___

7) $5x - 14y = -23$      $x = $ ___
   $-18x + 21y = 24$      $y = $ ___

8) $15x - 21y = -6$      $x = $ ___
   $2x - 3y = -2$      $y = $ ___

9) $-x + 3y = 3$      $x = $ ___
   $-14x + 16y = -10$      $y = $ ___

10) $x + 5y = 50$      $x = $ ___
    $3x + 10y = 80$      $y = $ ___

11) $6x - 7y = -8$      $x = $ ___
    $-x - 4y = -9$      $y = $ ___

12) $2x + 4y = -10$      $x = $ ___
    $2x - 8y = 14$      $y = $ ___

13) $4x + 3y = 12$      $x = $ ___
    $5x - 3y = 15$      $y = $ ___

14) $3x - 2y = 3$      $x = $ ___
    $7x - 8y = 22$      $y = $ ___

15) $3x + 2y = 5$      $x = $ ___
    $-10x - 4y = -14$      $y = $ ___

16) $10x + 7y = 1$      $x = $ ___
    $-5x - 7y = 24$      $y = $ ___

## Systems of Equations Word Problems

✎ **Find the answer for each word problem.**

1) Tickets to a movie cost $4 for adults and $3 for students. A group of friends purchased 8 tickets for $31.00. How many adults ticket did they buy? ____

2) At a store, Eva bought two shirts and five hats for $77.00. Nicole bought three same shirts and four same hats for $84.00. What is the price of each shirt? _____

3) A farmhouse shelters 18 animals, some are pigs, and some are ducks. Altogether there are 66 legs. How many pigs are there? _____

4) A class of 214 students went on a field trip. They took 36 vehicles, some cars and some buses. If each car holds 5 students and each bus hold 22 students, how many buses did they take? _____

5) A theater is selling tickets for a performance. Mr. Smith purchased 5 senior tickets and 3 child tickets for $105 for his friends and family. Mr. Jackson purchased 3 senior tickets and 5 child tickets for $79. What is the price of a senior ticket? $_____

6) The difference of two numbers is 10. Their sum is 20. What is the bigger number? $_____

7) The sum of the digits of a certain two–digit number is 7. Reversing its digits increase the number by 9. What is the number? _____

8) The difference of two numbers is 11. Their sum is 25. What are the numbers? _____

9) The length of a rectangle is 5 meters greater than 2 times the width. The perimeter of rectangle is 28 meters. What is the length of the rectangle? _____

10) Jim has 25 nickels and dimes totaling $1.80. How many nickels does he have? _____

# Answers of Worksheets

**One–Step Equations**

| | | | |
|---|---|---|---|
| 1) 30 | 9) 17 | 17) −14 | 25) 22 |
| 2) 7 | 10) −4 | 18) 20 | 26) 9 |
| 3) 4 | 11) 12 | 19) 45 | 27) 60 |
| 4) 6 | 12) 16 | 20) −13 | 28) 35 |
| 5) 5 | 13) 34 | 21) 34 | 29) 54 |
| 6) 11 | 14) −15 | 22) −5 | 30) 24 |
| 7) 11 | 15) −18 | 23) 47 | 31) 30 |
| 8) 8 | 16) 14 | 24) −42 | 32) −21 |

**Multi–Step Equations**

| | | | |
|---|---|---|---|
| 1) 2 | 9) 2 | 17) −2 | 25) 3 |
| 2) −7 | 10) −4 | 18) −7 | 26) 14 |
| 3) 4 | 11) −4 | 19) −5 | 27) 7 |
| 4) 8 | 12) −5 | 20) −11 | 28) 27 |
| 5) 6 | 13) −7 | 21) −4 | 29) −5 |
| 6) 2 | 14) 3 | 22) −2 | 30) 17 |
| 7) 10 | 15) −2 | 23) −7 | 31) −6 |
| 8) 2 | 16) 15 | 24) 6 | 32) −4 |

**Graphing Single–Variable Inequalities**

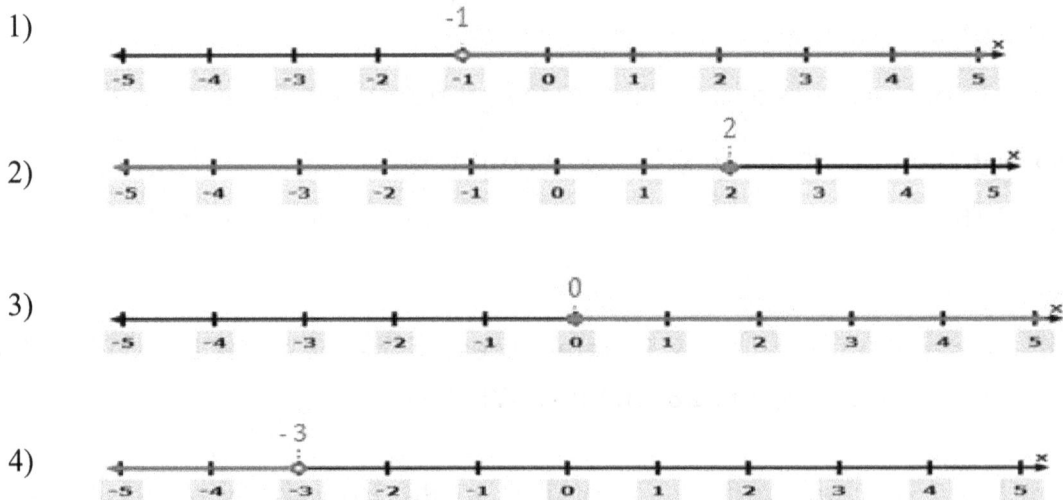

1)

2)

3)

4)

5)
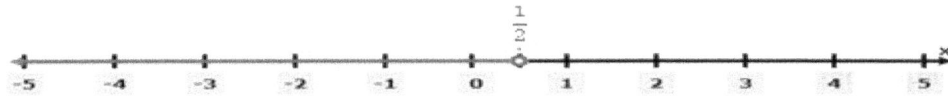

6)
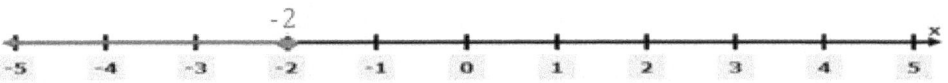

7)
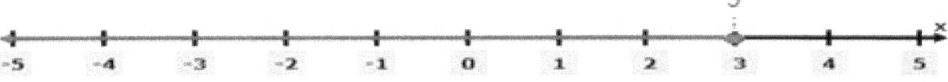

8)

## One–Step Inequalities

1)
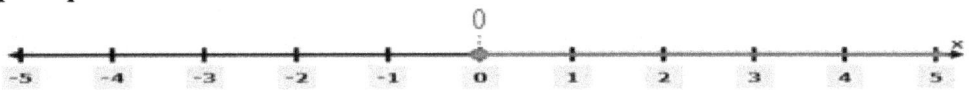

2)

3)
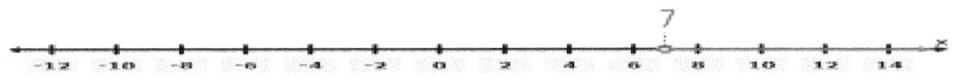

4)
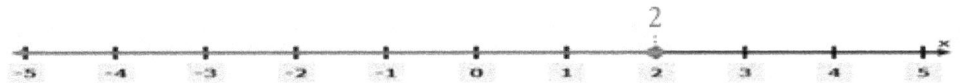

5)
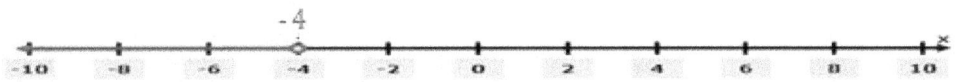

6)
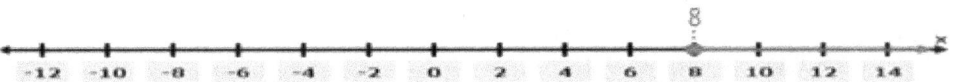

7)
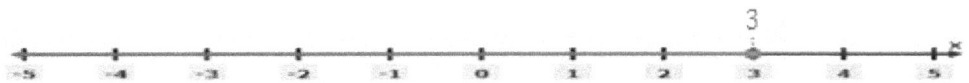

8)

## Multi-Step Inequalities

1) $x \leq 10$

2) $x \geq 0$

3) $x \leq 6$

4) $x \geq 3$

5) $x \geq 8$

6) $x \leq 4$

7) $x \leq 2$

8) $x \leq \frac{23}{6}$

9) $x \leq 8$

10) $x \leq 7$

11) $x < 10$

12) $x < 3$

13) $x \leq 5$

14) $x < 2$

15) $x \geq 11$

16) $x < 2$

17) $x < 3$

18) $x \leq 4$

19) $x > -6$

20) $x \leq 4$

21) $x < 0$

22) $x \leq 17$

23) $x \leq 4$

24) $x > 7$

25) $x < 6$        26) $x < 7$        27) $x > -2$        28) $x < -40$

**Systems of Equations**

1) $x = -1, y = 1$        7) $x = 1, y = 2$        13) $x = 3, y = 0$

2) $x = 0, y = -3$        8) $x = 8, y = 6$        14) $x = -2, y = -\frac{9}{2}$

3) $x = 7$        9) $x = 3, y = 2$        15) $x = 1, y = 1$

4) $x = 1, y = 2$        10) $x = -20, y = 14$        16) $x = 5, y = -7$

5) $x = -4, y = -3$        11) $x = 1, y = 2$

6) $x = 1, y = -2$        12) $x = -1, y = -2$

**Systems of Equations Word Problems**

1) 7        5) $18        9) 11 meters

2) $16        6) 15        10) 14

3) 15        7) 34

4) 2        8) 18, 7

# Chapter 7 :

# Linear Functions

## Topics that you'll practice in this chapter:

✓ Finding Slope

✓ Graphing Lines Using Line Equation

✓ Writing Linear Equations

✓ Graphing Linear Inequalities

✓ Finding Midpoint

✓ Finding Distance of Two Points

*Life is not linear; you have ups and downs. It's how you deal with the troughs that defines you.*

**Michael Lee-Chin**

# Finding Slope

✏ **Find the slope of each line.**

1) $y = x + 8$

2) $y = -3x + 5$

3) $y = 2x + 12$

4) $y = -4x + 19$

5) $y = 11 + 6x$

6) $y = 7 - 5x$

7) $y = 8x + 19$

8) $y = -9x + 20$

9) $y = -7x + 4$

10) $y = 3x - 8$

11) $y = \frac{1}{3}x + 8$

12) $y = -\frac{4}{5}x + 9$

13) $-3x + 6y = 30$

14) $4x + 4y = 16$

15) $3y - x = 10$

16) $8y - x = 5$

✏ **Find the slope of the line through each pair of points.**

17) $(2, 3), (7, 10)$

18) $(-3, 5), (2, 15)$

19) $(5, -3), (1, 9)$

20) $(-5, -5), (10, 25)$

21) $(22, 3), (7, 18)$

22) $(-16, 8), (-7, 26)$

23) $(25, 11), (29, 19)$

24) $(26, -19), (14, 17)$

25) $(22, -13), (20, -11)$

26) $(19, 7), (15, -3)$

27) $(5, 7), (11, 19)$

28) $(52, -62), (40, 70)$

# Graphing Lines Using Line Equation

✎ **Sketch the graph of each line.**

1) $y = x - 2$

2) $y = -3x + 2$

3) $x + y = 0$

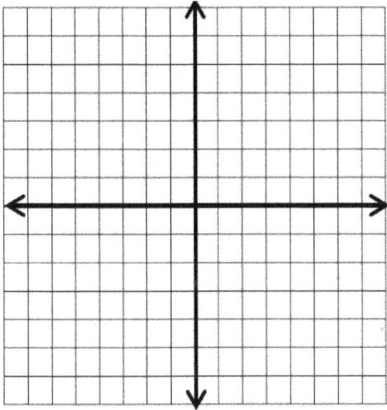
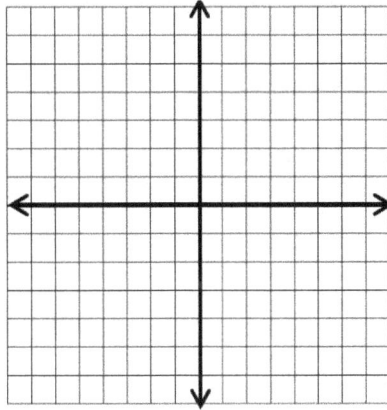
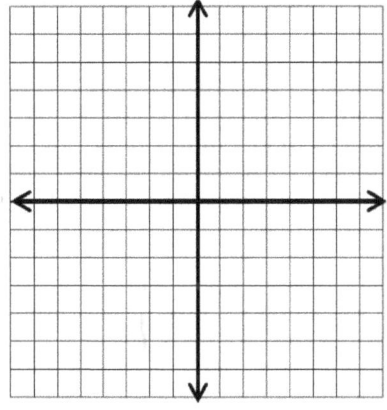

4) $x + y = -3$

5) $2x + 3y = -4$

6) $y - 3x + 6 = 0$

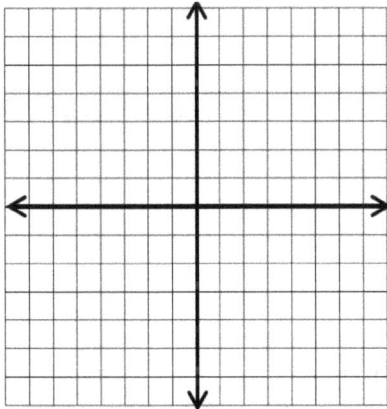
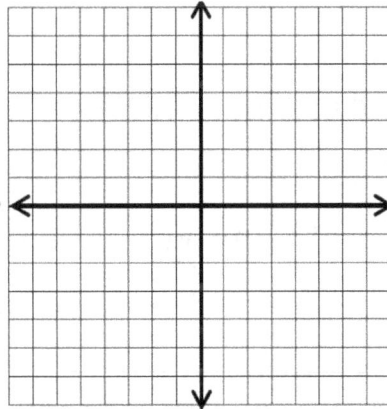
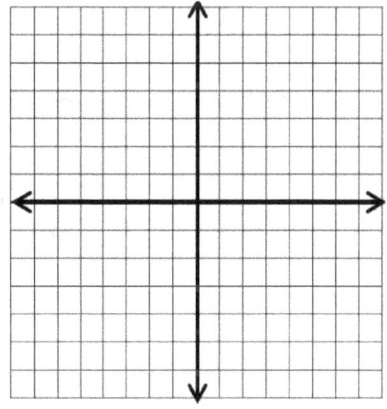

# Writing Linear Equations

✎ **Write the equation of the line through the given points.**

1) Through: $(2, -5), (3, 9)$

2) Through: $(-6, 3), (3, 12)$

3) Through: $(10, 7), (5, 27)$

4) Through: $(15, 11), (3, -1)$

5) Through: $(24, 17), (12, -7)$

6) Through: $(8, 29), (4, -7)$

7) Through: $(20, -16), (12, 0)$

8) Through: $(-3, 10), (2, -5)$

9) Through: $(-6, 17), (4, -3)$

10) Through: $(-8, 22), (5, -4)$

11) Through: $(9, 27), (3, -3)$

12) Through: $(11, 32), (9, 4)$

13) Through: $(-3, 13), (-4, 0)$

14) Through: $(-5, 5), (5, 15)$

15) Through: $(18, -32), (11, 3)$

16) Through: $(-4, 25), (4, -15)$

✎ **Find the answer for each problem.**

17) What is the equation of a line with slope 6 and intercept 12?
_____

18) What is the equation of a line with slope $-11$ and intercept $-4$?
_____

19) What is the equation of a line with slope $-3$ and passes through point $(5, 2)$? _____

20) What is the equation of a line with slope $-5$ and passes through point $(-2, -1)$? _____

21) The slope of a line is $-10$ and it passes through point $(-3, 0)$. What is the equation of the line? _____

22) The slope of a line is 8 and it passes through point $(0, 7)$. What is the equation of the line? _____

# Graphing Linear Inequalities

🖎 **Sketch the graph of each linear inequality.**

1) $y > 4x - 5$ 　　　 2) $y < 2x + 4$ 　　　 3) $y \leq -5x - 2$

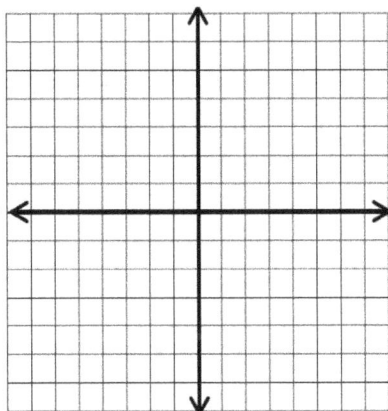 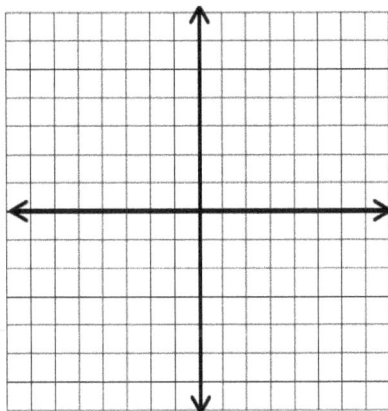 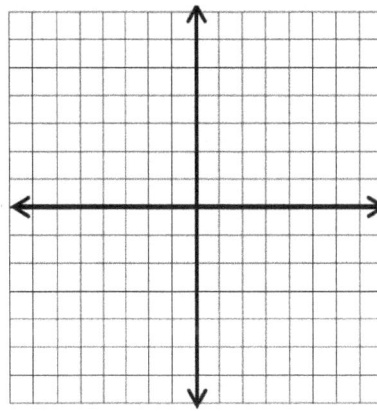

4) $4y \geq 12 + 4x$ 　　 5) $-12y < 3x - 24$ 　　 6) $5y \geq -15x + 10$

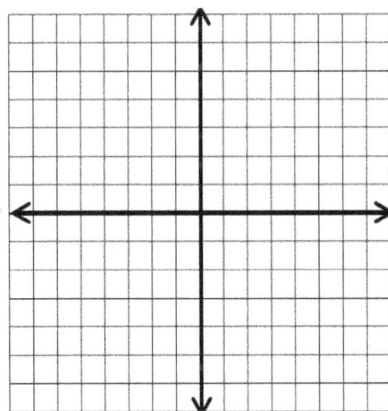 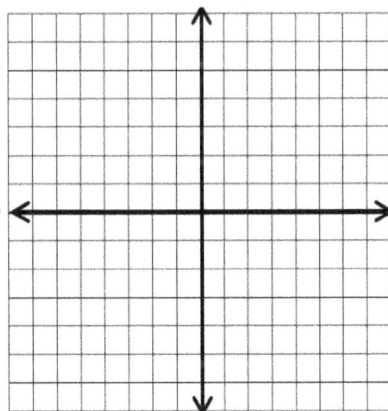 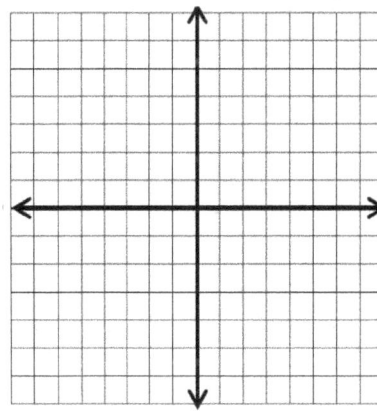

# Finding Midpoint

✍ **Find the midpoint of the line segment with the given endpoints.**

1) $(-4, -3), (2, 3)$

2) $(9, 0), (-1, 8)$

3) $(9, -6), (3, 14)$

4) $(-10, -6), (0, 8)$

5) $(2, -5), (14, -15)$

6) $(-10, -3), (4, -13)$

7) $(8, 7), (-8, 13)$

8) $(-3, 6), (-9, 2)$

9) $(-4, 5), (16, -9)$

10) $(7, 14), (9, -2)$

11) $(-8, 6), (6, 6)$

12) $(10, 5), (-2, -3)$

13) $(-5, 12), (-3, 3)$

14) $(12, 7), (8, -2)$

15) $(10, 2), (-6, 14)$

16) $(-1, -2), (-7, 10)$

17) $(7, -7), (13, -13)$

18) $(-3, -8), (11, -4)$

19) $(5, -11), (-8, 9)$

20) $(14, -4), (16, 14)$

21) $(0, -5), (8, -1)$

22) $(3, 0), (-21, 18)$

23) $(17, -3), (-7, -5)$

24) $(26, -12), (6, 24)$

✍ **Find the answer for each problem.**

25) One endpoint of a line segment is $(-3, 7)$ and the midpoint of the line segment is $(-6, 9)$. What is the other endpoint? _____

26) One endpoint of a line segment is $(-3, 7)$ and the midpoint of the line segment is $(1, 5)$. What is the other endpoint? _____

27) One endpoint of a line segment is $(-10, -16)$ and the midpoint of the line segment is $(2, 9)$. What is the other endpoint? _____

# Finding Distance of Two Points

✎ **Find the distance between each pair of points.**

1) $(6, 3), (-3, -9)$

2) $(5, 2), (-10, -6)$

3) $(8, 5), (8, 3)$

4) $(-8, -2), (2, 22)$

5) $(6, -7), (-3, -7)$

6) $(12, 0), (-9, -20)$

7) $(3, 20), (3, -5)$

8) $(10, 17), (5, 5)$

9) $(7, -2), (-4, -2)$

10) $(13, 4), (5, -2)$

11) $(11, 13), (5, 5)$

12) $(1, 4), (-23, -3)$

13) $(9, 8), (5, -4)$

14) $(-11, -4), (5, 8)$

15) $(-2, -6), (-2, -12)$

16) $(-1, -4), (23, 3)$

17) $(19, 3), (7, -6)$

18) $(-5, -2), (3, 4)$

19) $(2, 6), (2, -12)$

20) $(-4, -2), (8, -2)$

✎ **Find the answer for each problem.**

21) Triangle ABC is a right triangle on the coordinate system and its vertices are $(-2, 5)$, $(-2, 1)$, and $(1, 1)$. What is the area of triangle ABC? _____

22) Three vertices of a triangle on a coordinate system are $(3, -6)$, $(-5, -12)$, and $(3, -18)$. What is the perimeter of the triangle? _____

23) Four vertices of a rectangle on a coordinate system are $(-2, 2)$, $(-2, 6)$, $(4, 2)$, and $(4, 6)$. What is its perimeter? _____

# Answers of Worksheets

### Finding Slope

1) 1

2) −3

3) 2

4) −4

5) 6

6) −5

7) 8

8) −9

9) −7

10) 3

11) $\frac{1}{3}$

12) $-\frac{4}{5}$

13) $\frac{1}{2}$

14) −1

15) $\frac{1}{3}$

16) $\frac{1}{8}$

17) $\frac{7}{5}$

18) 2

19) −3

20) 2

21) −1

22) 2

23) 2

24) −3

25) −1

26) $\frac{5}{2}$

27) 2

28) −11

### Graphing Lines Using Line Equation

1) $y = x - 2$

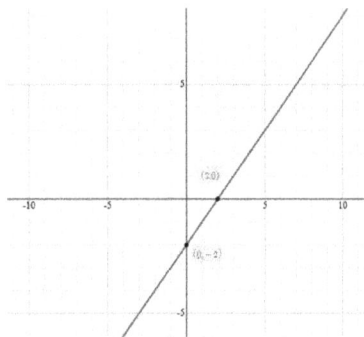

2) $y = -3x + 2$

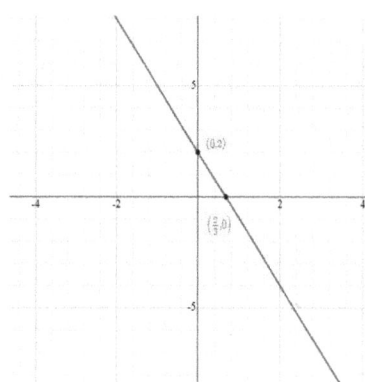

3) $x + y = 0$

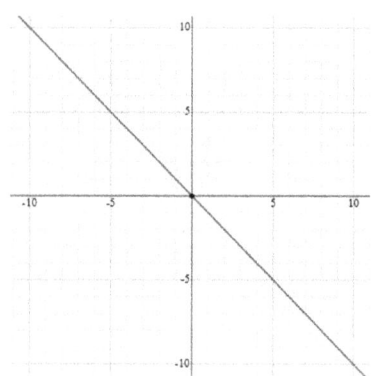

4) $x + y = -3$

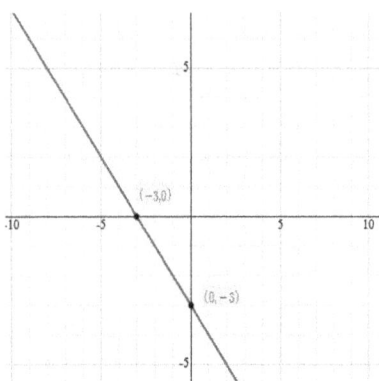

5) $2x + 3y = -4$

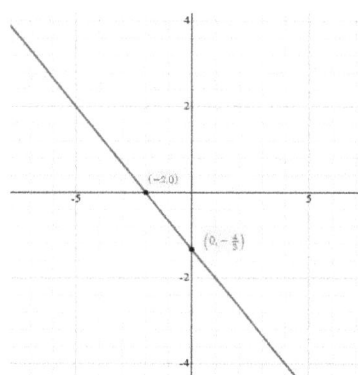

6) $y - 3x + 6 = 0$

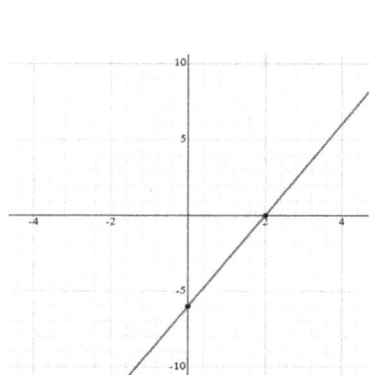

# PSAT Subject Test – Mathematics

## Writing Linear Equations

1) $y = 14x - 33$

2) $y = x + 9$

3) $y = -4x + 47$

4) $y = x - 4$

5) $y = 2x - 31$

6) $y = 9x - 43$

7) $y = -2x + 24$

8) $y = -3x + 1$

9) $y = -2x + 5$

10) $y = -2x + 6$

11) $y = 5x - 18$

12) $y = 14x - 122$

13) $y = 13x + 52$

14) $y = x + 10$

15) $y = -5x + 58$

16) $y = -5x + 5$

17) $y = 6x + 12$

18) $y = -11x - 4$

19) $y = -3x + 17$

20) $y = -5x - 11$

21) $y = -10x - 30$

22) $y = 8x + 7$

## Graphing Linear Inequalities

1) $y > 4x - 5$

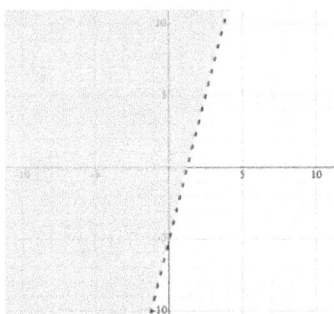

2) $y < 2x + 4$

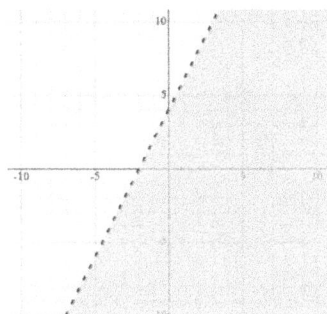

3) $y \leq -5x - 2$

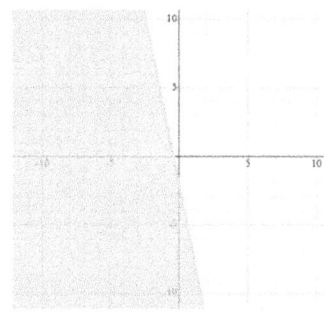

4) $4y \geq 12 + 4x$

5) $-12y < 3x - 24$

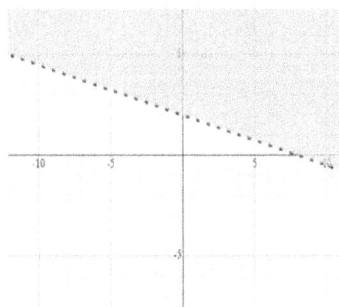

6) $5y \geq -15x + 10$

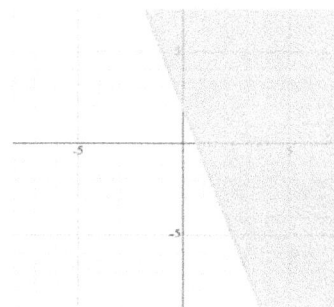

## Finding Midpoint

1) $(-1, 0)$

2) $(4, 4)$

3) $(6, 4)$

4) $(-5, 1)$

5) $(8, -10)$

6) $(-3, -8)$

7) $(0, 10)$

8) $(-6, 4)$

9) $(6, -2)$

10) $(8, 6)$

11) $(-1, 6)$

12) $(4, 1)$

13) $(-4, 7.5)$

14) $(10, 2.5)$

15) $(2, 8)$

16) $(-4, 4)$

17) $(10, -10)$

18) $(4, -6)$

19) $(-1.5, -1)$

20) $(15, 5)$

21) $(4, -3)$

22) $(-9, 9)$

23) $(5, -4)$

24) $(16, 6)$

25) $(-9, 11)$

26) $(5, 3)$

27) $(14, 34)$

**Finding Distance of Two Points**

1) 15

2) 17

3) 2

4) 26

5) 9

6) 29

7) 25

8) 13

9) 11

10) 10

11) 10

12) 25

13) $4\sqrt{10}$

14) 20

15) 6

16) 25

17) 15

18) 10

19) 18

20) 12

21) 6 *square units*

22) 32 *units*

23) 20 *units*

# Chapter 8 :

# Polynomials

## Topics that you'll practice in this chapter:

- ✓ Writing Polynomials in Standard Form
- ✓ Simplifying Polynomials
- ✓ Adding and Subtracting Polynomials
- ✓ Multiplying Monomials
- ✓ Multiplying and Dividing Monomials
- ✓ Multiplying a Polynomial and a Monomial
- ✓ Multiplying Binomials
- ✓ Factoring Trinomials
- ✓ Operations with Polynomials

*Mathematics is the supreme judge; from its decisions there is no appeal.*

*– Tobias Dantzig*

# Writing Polynomials in Standard Form

✎ **Write each polynomial in standard form.**

1) $11x - 7x =$

2) $-5 + 19x - 19x =$

3) $6x^5 - 12x^3 =$

4) $12 + 17x^4 - 12 =$

5) $5x^2 + 4x - 9x^3 =$

6) $-3x^2 + 12x^5 =$

7) $5x + 8x^3 - 2x^8 =$

8) $-7x^3 + 4x - 9x^6 =$

9) $3x^2 + 22 - 6x =$

10) $3 - 4x + 9x^4 =$

11) $13x^2 + 28x - 8x^3 =$

12) $16 + 4x^2 - 2x^3 =$

13) $19x^2 - 9x + 9x^4 =$

14) $3x^4 - 7x^2 - 2x^3 =$

15) $-51 + 3x^2 - 8x^4 =$

16) $7x^2 - 8x^6 + 4x^4 - 15 =$

17) $6x^4 - 4x^5 + 16 - 3x^3 =$

18) $-2x^6 + 4x - 7x^2 - 5x =$

19) $11x^7 + 8x^5 - 5x^7 - 3x^2 =$

20) $2x^2 - 12x^5 + 8x^2 + 3x^6 =$

21) $4x^5 - 11x^7 - 6x^3 + 16x^5 =$

22) $6x^3 + 3x^5 + 34x^4 - 8x^5 =$

23) $3x(4x + 5 - 2x^2) =$

24) $12x(x^6 + 4x^3) =$

25) $5x(3x^2 + 6x + 4) =$

26) $7x(4 - 2x + 6x^5) =$

27) $3x(4x^4 - 4x^3 + 2) =$

28) $4x(2x^5 + 6x^2 - 3) =$

29) $5x(3x^4 + 4x^3 + 2x) =$

30) $2x(3x - 2x^3 + 4x^6) =$

# Simplifying Polynomials

✎ **Simplify each expression.**

1) $3(4x - 20) =$

2) $5x(3x - 4) =$

3) $6x(5x - 7) =$

4) $3x(7x + 5) =$

5) $5x(4x - 3) =$

6) $6x(8x + 2) =$

7) $(3x - 2)(x - 4) =$

8) $(x - 5)(2x + 6) =$

9) $(x - 3)(x - 7) =$

10) $(3x + 4)(3x - 4) =$

11) $(5x - 4)(5x - 2) =$

12) $6x^2 + 6x^2 - 8x^4 =$

13) $3x - 2x^2 + 5x^3 + 7 =$

14) $7x + 4x^2 - 10x^3 =$

15) $12x^2 + 5x^5 - 6x^3 =$

16) $-5x^2 + 4x^6 + 6x^8 =$

17) $-12x^3 + 10x^5 - 4x^6 + 4x =$

18) $11 - 7x^2 + 4x^2 - 16x^3 + 11 =$

19) $2x^2 - 9x + 4x^3 + 15x - 10x =$

20) $13 - 7x^5 + 6x^5 - 4x^2 + 5 =$

21) $-5x^8 + x^6 - 14x^3 + 5x^8 =$

22) $(7x^4 - 4) + (7x^4 - 2x^4) =$

23) $3(3x^4 - 4x^3 - 6x^4) =$

24) $-5(x^9 + 8) - 5(10 - x^9) =$

25) $8x^3 - 9x^4 - 2x + 19 - 8x^3 =$

26) $11 - 8x^3 + 6x^3 - 7x^5 + 6 =$

27) $(5x^3 - 4x) - (6x - 2 - 6x^3) =$

28) $4x^2 - 5x^4 - x(3x^3 + 2x) =$

29) $6x + 6x^5 - 10 - 4(x^5 - 3) =$

30) $4 - 3x^4 + (6x^5 - 2x^4 + 5x^5) =$

31) $-(x^5 + 4) - 8(3 + x^5) =$

32) $(4x^3 - 3x) - (3x - 5x^3) =$

# Adding and Subtracting Polynomials

✎ Add or subtract expressions.

1) $(-2x^2 - 3) + (3x^2 + 4) =$

2) $(4x^3 + 6) - (7 - 2x^3) =$

3) $(4x^5 + 5x^2) - (2x^5 + 15) =$

4) $(6x^3 - 2x^2) + (5x^2 - 4x) =$

5) $(10x^4 + 28x) - (34x^4 + 6) =$

6) $(7x^2 - 3) + (7x^2 + 3) =$

7) $(9x^2 + 4) - (10 - 5x^2) =$

8) $(6x^2 + x^5) - (x^5 + 4) =$

9) $(4x^3 - x) + (3x - 7x^3) =$

10) $(11x + 10) - (8x + 10) =$

11) $(15x^3 - 3x) - (3x - 4x^3) =$

12) $(4x - x^5) - (6x^5 + 8x) =$

13) $(2x^2 - 7x^7) - (4x^7 - 6x) =$

14) $(3x^2 - 5) + (8x^2 + 4x^5) =$

15) $(9x^4 + 5x^5) - (x^5 - 9x^4) =$

16) $(-4x^3 - 2x) + (9x - 5x^3) =$

17) $(4x - 3x^2) - (148x^2 + x) =$

18) $(5x - 8x^4) - (3x^4 - 4x^2) =$

19) $(8x^4 - 4) + (2x^4 - 3x^2) =$

20) $(5x^6 + 7x^3) - (x^3 - 5x^6) =$

21) $(-2x^2 + 20x^5 + 5x^4) + (12x^4 + 8x^5 + 24x^2) =$

22) $(7x^4 - 9x^7 - 6x) - (-3x^4 - 9x^7 + 6x) =$

23) $(14x + 12x^4 - 18x^6) + (20x^4 + 18x^6 - 10x) =$

24) $(5x^8 - 6x^6 - 4x) - (5x^3 + 9x^6 - 7x) =$

25) $(11x^2 - 6x^4 - 3x) - (-4x^2 - 12x^4 + 9x) =$

26) $(-5x^9 + 14x^3 + 3x^7) + (10x^7 + 26x^3 + 3x^9) =$

# Multiplying Monomials

✎ **Simplify each expression.**

1) $6u^8 \times (-u^2) =$

2) $(-5p^8) \times (-2p^3) =$

3) $4xy^3z^5 \times 3z^4 =$

4) $3u^5t \times 8ut^4 =$

5) $(-5a^2) \times (-7a^3b^6) =$

6) $-3a^4b^3 \times 6a^2b =$

7) $13xy^5 \times x^4y^4 =$

8) $6p^4q^3 \times (-8pq^6) =$

9) $8s^4t^3 \times 4st^3 =$

10) $(-6x^4y^3) \times 6x^2y =$

11) $3xy^7z \times 12z^3 =$

12) $24xy \times x^2y =$

13) $13pq^4 \times (-3p^2q) =$

14) $13s^3t^4 \times st^4 =$

15) $11p^5 \times (-6p^3) =$

16) $(-8p^3q^5r) \times 3pq^4r^6 =$

17) $(-4a^4) \times (-7a^3b) =$

18) $6u^6v^2 \times (-5u^3v^4) =$

19) $9u^5 \times (-3u) =$

20) $-6xy^5 \times 4x^2y =$

21) $13y^5z^3 \times (-y^3z) =$

22) $8a^4bc^3 \times 2abc^3 =$

23) $(-7p^5q^6) \times (-5p^4q^2) =$

24) $4u^5v^3 \times (-4u^7v^3) =$

25) $17y^4z^5 \times (-y^6z) =$

26) $(-5pq^3r^2) \times 8p^2q^4r =$

27) $3ab^5c^6 \times 5a^4bc^2 =$

28) $6x^3yz^2 \times 3x^2y^7z^3 =$

# Multiplying and Dividing Monomials

✎ **Simplify each expression.**

1) $(5x^5)(2x^2) =$

2) $(4x^4)(6x^2) =$

3) $(3x^4)(7x^4) =$

4) $(5x^6)(4x^2) =$

5) $(12x^4)(3x^6) =$

6) $(4yx^8)(8y^4x^3) =$

7) $(14x^4y)(x^3y^5) =$

8) $(-5x^3y^4)(2x^3y^5) =$

9) $(-6x^4y^2)(-3x^3y^5) =$

10) $(5x^3y)(-5x^2y^3) =$

11) $(6x^4y^3)(4x^3y^4) =$

12) $(4x^3y^2)(5x^2y^4) =$

13) $(12x^3y^6)(4x^4y^{10}) =$

14) $(15x^3y^5)(3x^4y^6) =$

15) $(7x^2y^7)(8x^6y^7) =$

16) $(-3x^3y^8)(7x^9y^4) =$

17) $\dfrac{5x^6y^6}{xy^4} =$

18) $\dfrac{19x^7y^5}{19x^6y} =$

19) $\dfrac{56x^4y^4}{8xy} =$

20) $\dfrac{81x^5y^6}{9x^4y^5} =$

21) $\dfrac{36x^7y^6}{9x^2y^3} =$

22) $\dfrac{48x^9y^7}{4x^4y^6} =$

23) $\dfrac{88x^{18}y^{12}}{11x^8y^9} =$

24) $\dfrac{30x^7y^6}{6x^8y^3} =$

25) $\dfrac{150x^7y^6}{30x^4y^6} =$

26) $\dfrac{-42x^{18}y^{14}}{6x^4y^9} =$

27) $\dfrac{-36x^7y^8}{9x^5y^8} =$

## Multiplying a Polynomial and a Monomial

✎ **Find each product.**

1) $x(2x + 4) =$

2) $6(4 - 2x) =$

3) $5x(4x + 2) =$

4) $x(-4x + 5) =$

5) $8x(2x - 2) =$

6) $6(2x - 4y) =$

7) $7x(5x - 5) =$

8) $3x(12x + 2y) =$

9) $4x(x + 6y) =$

10) $11x(3x + 4y) =$

11) $7x(3x + 2) =$

12) $10x(4x - 10y) =$

13) $9x(3x - 2y) =$

14) $7x(x - 4y + 6) =$

15) $8x(2x^2 + 5y^2) =$

16) $12x(2x + 3y) =$

17) $4(2x^4 - 4y^4) =$

18) $4x(-3x^2y + 4y) =$

19) $-4(5x^3 - 2xy + 4) =$

20) $4(x^2 - 5xy - 6) =$

21) $8x(2x^3 - 5xy + 2x) =$

22) $-6x(-2x^3 - 6x + 2xy) =$

23) $3(2x^2 + xy - 9y^2) =$

24) $4x(5x^3 - 3x + 7) =$

25) $6(3x^{22} - 2x - 5) =$

26) $x^2(-2x^3 + 4x + 3) =$

27) $x^2(4x^3 + 10 - 2x) =$

28) $4x^4(3x^3 - 2x + 5) =$

29) $2x^2(4x^4 - 5xy + 7y^3) =$

30) $5x^2(5x^4 - 3x + 9) =$

31) $7x^2(6x^2 + 3x - 6) =$

32) $4x(x^3 - 4xy + 2y^2) =$

## Multiplying Binomials

✍ **Find each product.**

1) $(x + 3)(x + 6) =$

2) $(x - 4)(x + 3) =$

3) $(x - 3)(x - 8) =$

4) $(x + 8)(x + 9) =$

5) $(x - 2)(x - 12) =$

6) $(x + 5)(x + 5) =$

7) $(x - 6)(x + 7) =$

8) $(x - 8)(x - 3) =$

9) $(x + 7)(x + 12) =$

10) $(x - 4)(x + 8) =$

11) $(x + 8)(x + 8) =$

12) $(x + 2)(x + 7) =$

13) $(x - 6)(x + 6) =$

14) $(x - 5)(x + 5) =$

15) $(x + 11)(x + 11) =$

16) $(x + 6)(x + 9) =$

17) $(x - 2)(x + 2) =$

18) $(x - 4)(x + 7) =$

19) $(3x + 5)(x + 6) =$

20) $(5x - 6)(4x + 8) =$

21) $(x - 7)(3x + 7) =$

22) $(x - 9)(x - 4) =$

23) $(x - 12)(x + 2) =$

24) $(2x - 4)(5x + 4) =$

25) $(3x - 8)(x + 8) =$

26) $(7x - 2)(6x + 3) =$

27) $(4x + 5)(3x + 5) =$

28) $(7x - 4)(9x + 4) =$

29) $(x + 2)(2x - 8) =$

30) $(5x - 4)(5x + 4) =$

31) $(3x + 2)(3x - 7) =$

32) $(x^2 + 8)(x^2 - 8) =$

# Factoring Trinomials

🖎 **Factor each trinomial.**

1) $x^2 + 8x + 12 =$

2) $x^2 - 6x + 5 =$

3) $x^2 + 15x + 36 =$

4) $x^2 - 12x + 35 =$

5) $x^2 - 11x + 18 =$

6) $x^2 - 9x + 18 =$

7) $x^2 + 18x + 72 =$

8) $x^2 - x - 72 =$

9) $x^2 + 4x - 21 =$

10) $x^2 - 13x + 22 =$

11) $x^2 + 2x - 24 =$

12) $x^2 - 3x - 40 =$

13) $x^2 - 3x - 70 =$

14) $x^2 + 26x + 169 =$

15) $4x^2 - 7x - 15 =$

16) $x^2 - 14x + 33 =$

17) $10x^2 + 5x - 15 =$

18) $6x^2 - 4x - 42 =$

19) $x^2 + 12x + 36 =$

20) $5x^2 + 17x - 12 =$

🖎 **Calculate each problem.**

21) The area of a rectangle is $x^2 - x - 56$. If the width of rectangle is $x + 7$, what is its length? _____

22) The area of a parallelogram is $4x^2 + 17x - 15$ and its height is $x + 5$. What is the base of the parallelogram? _____

23) The area of a rectangle is $6x^2 - 22x + 12$. If the width of the rectangle is $3x - 2$, what is its length? _____

## Operations with Polynomials

✏️ **Find each product.**

1) $4(5x + 3) =$ _____

2) $8(2x + 6) =$ _____

3) $2(5x - 2) =$ _____

4) $-4(7x - 3) =$ _____

5) $3x^2(9x + 1) =$ _____

6) $4x^6(7x - 9) =$ _____

7) $3x^4(-7x + 3) =$ _____

8) $-8x^4(5x - 8) =$ _____

9) $7(x^2 + 5x - 3) =$ _____

10) $9(5x^2 - 7x + 5) =$ _____

11) $3(3x^2 + 3x + 2) =$ _____

12) $5x(3x^2 + 5x + 8) =$ _____

13) $(5x + 7)(3x - 3) =$ _____

14) $(9x + 3)(3x - 5) =$ _____

15) $(6x + 3)(4x - 2) =$ _____

16) $(7x - 2)(3x + 5) =$ _____

✏️ **Calculate each problem.**

17) The measures of two sides of a triangle are $(2x + 5y)$ and $(6x - 3y)$. If the perimeter of the triangle is $(13x + 4y)$, what is the measure of the third side? _____

18) The height of a triangle is $(8x + 5)$ and its base is $(4x - 3)$. What is the area of the triangle? _____

19) One side of a square is $(6x + 2)$. What is the area of the square? _____

20) The length of a rectangle is $(5x - 8y)$ and its width is $(15x + 8y)$. What is the perimeter of the rectangle? _____

21) The side of a cube measures $(x + 2)$. What is the volume of the cube? _____

22) If the perimeter of a rectangle is $(28x + 6y)$ and its width is $(5x + 2y)$, what is the length of the rectangle? _____

# Answers of Worksheets

**Writing Polynomials in Standard Form**

1) $4x$

2) $-5$

3) $6x^5 - 12x^3$

4) $14x^4$

5) $-9x^3 + 5x^2 + 4x$

6) $12x^5 - 3x^2$

7) $-2x^8 + 8x^3 + 5x$

8) $-9x^6 - 7x^3 + 4x$

9) $3x^2 - 6x + 22$

10) $9x^4 - 4x + 3$

11) $-8x^3 + 13x^2 + 28x$

12) $-2x^3 + 4x^2 + 16$

13) $9x^4 + 19x^2 - 9x$

14) $3x^4 - 2x^3 - 7x^2$

15) $-8x^4 + 3x^2 - 51$

16) $-8x^6 + 4x^4 + 7x^2 - 15$

17) $-4x^5 + 6x^4 - 3x^3 + 16$

18) $-2x^6 - 7x^2 - x$

19) $6x^7 + 8x^5 - 3x^2$

20) $3x^6 - 12x^5 + 10x^2$

21) $-11x^7 + 20x^5 - 6x^3$

22) $-5x^5 + 34x^4 + 6x^3$

23) $-6x^3 + 12x^2 + 15x$

24) $12x^7 + 48x^4$

25) $15x^3 + 30x^2 + 20x$

26) $42x^6 - 14x^2 + 28x$

27) $12x^5 - 12x^4 + 6x$

28) $8x^6 + 24x^3 - 12x$

29) $15x^5 + 20x^4 + 10x^2$

30) $8x^7 - 4x^4 + 6x^2$

**Simplifying Polynomials**

1) $12x - 60$

2) $15x^2 - 20x$

3) $30x^2 - 42x$

4) $21x^2 + 15x$

5) $20x^2 - 15x$

6) $48x^2 + 12x$

7) $3x^2 - 14x + 8$

8) $2x^2 - 4x - 30$

9) $x^2 - 10x + 21$

10) $9x^2 - 16$

11) $25x^2 - 30x + 8$

12) $-8x^4 + 12x^2$

13) $5x^3 - 2x^2 + 3x + 7$

14) $-10x^3 + 4x^2 + 7x$

15) $5x^5 - 6x^3 + 12x^2$

16) $6x^8 + 4x^6 - 5x^2$

17) $-4x^6 + 10x^5 - 12x^3 + 4x$

18) $-16x^3 - 3x^2 + 22$

19) $4x^3 + 2x^2 - 4x$

20) $-x^5 - 4x^2 + 18$

21) $x^6 - 14x^3$

22) $12x^4 - 4$

23) $-9x^4 - 12x^3$

24) $-90$

25) $-9x^4 - 2x + 19$

26) $-7x^5 - 2x^3 + 17$

27) $11x^3 - 10x + 2$

28) $-8x^4 + 2x^2$

29) $2x^5 + 6x + 2$

30) $11x^5 - 5x^4 + 4$

31) $-9x^5 - 28$

32) $9x^3 - 6x$

## Adding and Subtracting Polynomials

1) $x^2 + 1$

2) $6x^3 - 1$

3) $2x^5 + 5x^2 - 15$

4) $6x^3 + 3x^2 - 4x$

5) $-24x^4 + 28x - 6$

6) $14x^2$

7) $14x^2 - 6$

8) $6x^2 - 4$

9) $-3x^3 + 2x$

10) $3x$

11) $19x^3 - 6x$

12) $-7x^5 - 4x$

13) $-11x^7 + 2x^2 + 6x$

14) $4x^5 + 11x^2 - 5$

15) $4x^5 + 18x^4$

16) $-9x^3 + 7x$

17) $-151x^2 + 3x$

18) $-11x^4 + 4x^2 + 5x$

19) $10x^4 - 3x^2 - 4$

20) $10x^6 + 6x^3$

21) $28x^5 + 17x^4 + 22x^2$

22) $10x^4 - 12x$

23) $32x^4 + 4x$

24) $5x^8 - 15x^6 - 5x^3 + 3x$

25) $6x^4 + 15x^2 - 12x$

26) $-2x^9 + 13x^7 + 40x^3$

## Multiplying Monomials

1) $-6u^{10}$

2) $10p^{11}$

3) $12xy^3z^9$

4) $24u^6t^5$

5) $35a^5b^6$

6) $-18a^6b^4$

7) $13x^5y^9$

8) $-48p^5q^9$

9) $32s^5t^6$

10) $-36x^6y^4$

11) $36xy^7z^4$

12) $24px^3y^2$

13) $-39p^3q^5$

14) $13s^4t^8$

15) $-66p^8$

16) $-24p^4q^9r^7$

17) $28a^7b$

18) $-30u^9v^6$

19) $-27u^6$

20) $-24x^3y^6$

21) $-13y^8z^4$

22) $16a^5b^2c^6$

23) $35p^9q^8$

24) $-16u^{12}v^6$

25) $-17y^{10}z^6$

26) $-40p^3q^7r^3$

27) $15a^5b^6c^8$

28) $18x^5y^8z^5$

## Multiplying and Dividing Monomials

1) $10x^7$

2) $24x^6$

3) $21x^8$

4) $20x^8$

5) $36x^{10}$

6) $32x^{11}y^5$

7) $14x^7y^6$

8) $-10x^6y^9$

9) $18x^7y^7$

10) $-25x^5y^4$

11) $24x^7y^7$

12) $20x^5y^6$

13) $48x^7y^{16}$

14) $45x^7y^{11}$

15) $56x^8y^{14}$

16) $-21x^{12}y^{12}$

17) $5x^5y^2$

18) $xy^4$

19) $7x^3y^3$

20) $9xy$

21) $4x^5y^3$

22) $12x^5y$

23) $8x^{10}y^3$

24) $5x^{-1}y^3$

25) $5x^3$

26) $-7x^{14}y^5$

27) $-4x^2$

## Multiplying a Polynomial and a Monomial

1) $2x^2 + 4x$

2) $-12x + 24$

3) $20x^2 + 10x$

4) $-4x^2 + 5x$

5) $16x^2 - 16x$

6) $12x - 24y$

7) $35x^2 - 35x$

8) $36x^2 + 6xy$

9) $4x^2 + 24xy$

10) $33x^2 + 44xy$

11) $21x^2 + 14x$

12) $40x^2 - 100xy$

13) $27x^2 - 18xy$

14) $7x^2 - 28xy + 42x$

15) $16x^3 + 40xy^2$

16) $24x^2 + 36xy$

17) $8x^4 - 16y^4$

18) $-12x^3y + 16xy$

19) $-20x^3 + 8xy - 16$

20) $4x^2 - 20xy - 24$

21) $16x^4 - 40x^2y + 16x^2$

22) $12x^4 + 36x^2 - 12x^2y$

23) $6x^2 + 3xy - 27y^2$

24) $20x^4 - 12x^2 + 28x$

25) $18x^{22} - 12x - 30$

26) $-2x^5 + 4x^3 + 3x^2$

27) $4x^5 - 2x^3 + 10x^2$

28) $12x^7 - 8x^5 + 20x^4$

29) $8x^6 - 10x^3y + 14x^2y^3$

30) $25x^6 - 15x^3 + 45x^2$

31) $42x^4 + 21x^3 - 42x^2$

32) $4x^4 - 16x^2y + 8xy^2$

## Multiplying Binomials

1) $x^2 + 9x + 18$

2) $x^2 - x - 12$

3) $x^2 - 11x + 24$

4) $x^2 + 17x + 72$

5) $x^2 - 14x + 24$

6) $x^2 + 10x + 25$

7) $x^2 + x - 42$

8) $x^2 - 11x + 24$

9) $x^2 + 19x + 84$

10) $x^2 + 4x - 32$

11) $x^2 + 16x + 64$

12) $x^2 + 9x + 14$

13) $x^2 - 36$

14) $x^2 - 25$

15) $x^2 + 22x + 121$

16) $x^2 + 15x + 54$

17) $x^2 - 4$

18) $x^2 + 3x - 28$

19) $3x^2 + 23x + 30$

20) $20x^2 + 16x - 48$

21) $3x^2 - 14x - 49$

22) $x^2 - 13x + 36$

23) $x^2 - 10x - 24$

24) $10x^2 - 12x - 16$

25) $3x^2 + 16x - 64$

26) $42x^2 + 9x - 6$

27) $12x^2 + 35x + 25$

28) $63x^2 - 8x - 16$

29) $2x^2 - 4x - 16$

30) $25x^2 - 16$

31) $9x^2 - 15x - 14$

32) $x^4 - 64$

**Factoring Trinomials**

1) $(x + 6)(x + 2)$

2) $(x - 5)(x - 1)$

3) $(x + 12)(x + 3)$

4) $(x - 5)(x - 7)$

5) $(x - 2)(x - 9)$

6) $(x - 6)(x - 3)$

7) $(x + 6)(x + 12)$

8) $(x + 8)(x - 9)$

9) $(x - 3)(x + 7)$

10) $(x - 11)(x - 2)$

11) $(x - 4)(x + 6)$

12) $(x - 8)(x + 5)$

13) $(x + 7)(x - 10)$

14) $(x + 13)(x + 13)$

15) $(4x + 5)(x - 3)$

16) $(x - 11)(x - 3)$

17) $(5x - 5)(2x + 3)$

18) $(2x - 6)(3x + 7)$

19) $(x + 6)(x + 6)$

20) $(5x - 3)(x + 4)$

21) $(x - 8)$

22) $(4x - 3)$

23) $(2x - 6)$

**Operations with Polynomials**

1) $20x + 12$

2) $16x + 48$

3) $10x - 4$

4) $-28x + 12$

5) $27x^3 + 3x^2$

6) $28x^7 - 36x^6$

7) $-21x^5 + 9x^4$

8) $-40x^5 + 64x^4$

9) $7x^2 + 35x - 21$

10) $45x^2 - 63x + 45$

11) $9x^2 + 9x + 6$

12) $15x^3 + 25x^2 + 40x$

13) $15x^2 + 6x - 21$

14) $27x^2 - 36x - 15$

15) $24x^2 - 6$

16) $21x^2 + 29x - 10$

17) $(5x + 2y)$

18) $16x^2 - 2x - \frac{15}{2}$

19) $36x^2 + 24x + 4$

20) $40x$

21) $x^3 + 6x^2 + 12x + 8$

22) $(9x + y)$

# Chapter 9 :

# Complex Numbers

## Topics that you'll practice in this chapter:

- ✓ Adding and Subtracting Complex Numbers
- ✓ Multiplying and Dividing Complex Numbers
- ✓ Graphing Complex Numbers
- ✓ Rationalizing Imaginary Denominators

*Mathematics is a hard thing to love. It has the unfortunate habit, like a rude dog, of turning its most unfavorable side towards you when you first make contact with it. — David Whiteland*

## Adding and Subtracting Complex Numbers

✎ **Simplify.**

1) $(7i) - (3i) =$

2) $(5i) + (4i) =$

3) $(2i) + (8i) =$

4) $(-8i) - (3i) =$

5) $(14i) + (6i) =$

6) $(6i) - (-10i) =$

7) $(-2i) + (-5i) =$

8) $(13i) - (5i) =$

9) $(-22i) - (11i) =$

10) $(-4i) + (2 + 6i) =$

11) $(10 - 5i) + (-3i) =$

12) $(-8i) + (6 + 12i) =$

13) $1 + (5 - 4i) =$

14) $(13i) - (-8 + 2i) =$

15) $(8 + 12i) - (-10i) =$

16) $(10 + i) + (-5i) =$

17) $(11i) - (-7i + 9) =$

18) $(10i + 12) + (-2i) =$

19) $(20) - (16 + 4i) =$

20) $(3 + 3i) + (8 + 4i) =$

21) $(15 - 7i) + (3 + 4i) =$

22) $(12 + 6i) + (10 + 17i) =$

23) $(-5 + 6i) - (-16 - 12i) =$

24) $(-4 + 14i) - (-9 + 11i) =$

25) $(-22 + 4i) - (-7 - 22i) =$

26) $(-26 - 18i) + (3 + 34i) =$

27) $(-19 - 13i) - (-7 - 20i) =$

28) $-21 + (5i) + (-32 + 14i) =$

29) $30 - (7i) + (3 - 11i) =$

30) $28 + (-32 - 10i) - 7 =$

31) $(-44i) + (2 - 7i) + 9 =$

32) $(-21i) - (12 - 9i) + 21i =$

## Multiplying and Dividing Complex Numbers

✎ **Simplify.**

1) $(5i)(-3i) =$

2) $(-8i)(2i) =$

3) $(3i)(-3i)(-3i) =$

4) $(6i)(-6i) =$

5) $(-3 - 4i)(2 + i) =$

6) $(5 - 2i)^2 =$

7) $(5 - 2i)(6 - 4i) =$

8) $(1 + 6i)^2 =$

9) $(5i)(-3i)(2 - 4i) =$

10) $(11 - 2i)(2 - 4i) =$

11) $(-3 + i)(6 + 5i) =$

12) $(2 - 8i)(6 - 4i) =$

13) $3(4i) - (6i)(-2 + 5i) =$

14) $\dfrac{5}{-25i} =$

15) $\dfrac{3-4i}{-5i} =$

16) $\dfrac{6+12i}{2i} =$

17) $\dfrac{20i}{-5+4i} =$

18) $\dfrac{-6-9i}{4i} =$

19) $\dfrac{4i}{8-2i} =$

20) $\dfrac{4-7i}{6-2i} =$

21) $\dfrac{3-2i}{-1-1i} =$

22) $\dfrac{-5-5i}{-4-i} =$

23) $\dfrac{-6+2i}{-10-4i} =$

24) $\dfrac{-8-4i}{-2+4i} =$

25) $\dfrac{2+3i}{1-4i} =$

# Graphing Complex Numbers

✎ **Identify each complex number graphed.**

1)

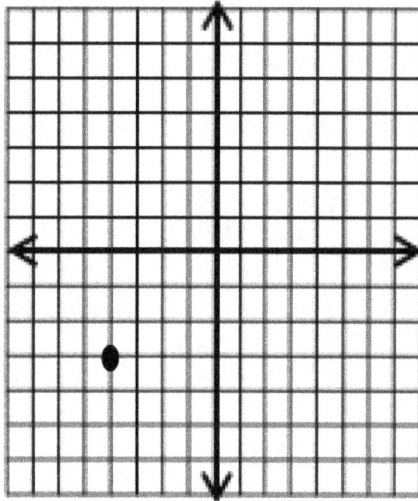

2)

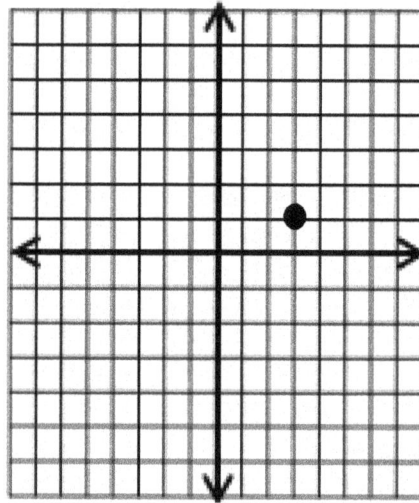

3)

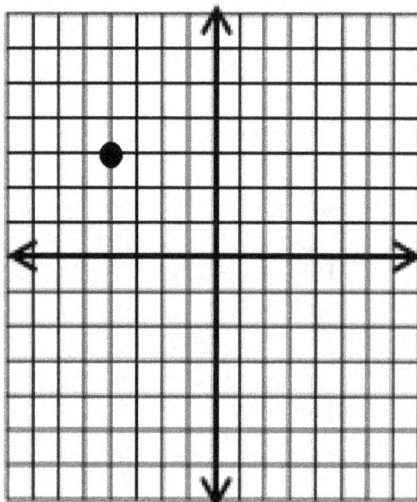

4)

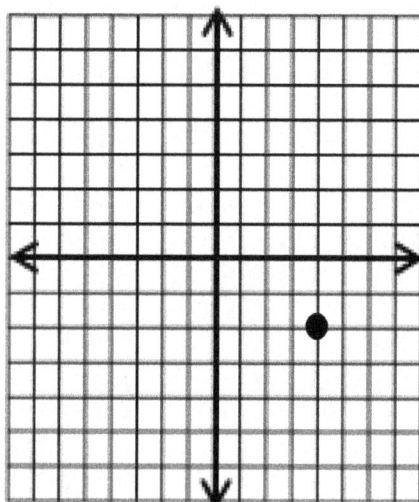

# Rationalizing Imaginary Denominators

✎ **Simplify.**

1) $\dfrac{-7}{-7i} =$

2) $\dfrac{-3}{-15i} =$

3) $\dfrac{-3}{-39i} =$

4) $\dfrac{24}{-3i} =$

5) $\dfrac{5}{2i} =$

6) $\dfrac{16}{-4i} =$

7) $\dfrac{14}{-6i} =$

8) $\dfrac{-17}{3i} =$

9) $\dfrac{4x}{5yi} =$

10) $\dfrac{10-10i}{-2i} =$

11) $\dfrac{5-11i}{-i} =$

12) $\dfrac{21+4i}{4i} =$

13) $\dfrac{8i}{-1+4i} =$

14) $\dfrac{10i}{-6+8i} =$

15) $\dfrac{-25-5i}{-5+5i} =$

16) $\dfrac{-7-2i}{3+1i} =$

17) $\dfrac{-12-6i}{8-6i} =$

18) $\dfrac{-14+7i}{-7i} =$

19) $\dfrac{12+3i}{3i} =$

20) $\dfrac{-2-i}{4-3i} =$

21) $\dfrac{-11+4i}{-5i} =$

22) $\dfrac{8+2i}{-5-2i} =$

23) $\dfrac{-9-5i}{-8-2i} =$

24) $\dfrac{4i-1}{-5-2i} =$

# Answers of Worksheets – Chapter 9

### Adding and Subtracting Complex Numbers

| | | | |
|---|---|---|---|
| 1) $4i$ | 9) $-33i$ | 17) $-9 + 18i$ | 25) $-15 + 26i$ |
| 2) $9i$ | 10) $2 + 2i$ | 18) $12 + 8i$ | 26) $-23 + 16i$ |
| 3) $10i$ | 11) $10 - 8i$ | 19) $4 - 4i$ | 27) $-12 + 7i$ |
| 4) $-11i$ | 12) $6 + 4i$ | 20) $11 + 7i$ | 28) $-53 + 19i$ |
| 5) $20i$ | 13) $6 - 4i$ | 21) $18 - 3i$ | 29) $33 - 18i$ |
| 6) $16i$ | 14) $8 + 11i$ | 22) $22 + 23i$ | 30) $-11 - 10i$ |
| 7) $-7i$ | 15) $8 + 22i$ | 23) $11 + 18i$ | 31) $11 - 51i$ |
| 8) $8i$ | 16) $10 - 4i$ | 24) $5 + 3i$ | 32) $-12 + 9i$ |

### Multiplying and Dividing Complex Numbers

| | | | |
|---|---|---|---|
| 1) $15$ | 9) $30 - 60i$ | 16) $-6 + 3i$ | 22) $\frac{25}{17} + \frac{15}{17}i$ |
| 2) $16$ | 10) $14 - 48i$ | 17) $\frac{80}{41} - \frac{100}{41}i$ | 23) $\frac{13}{29} - \frac{11}{29}i$ |
| 3) $-27i$ | 11) $-23 - 9i$ | 18) $\frac{9}{4} - \frac{3}{2}i$ | 24) $2i$ |
| 4) $36$ | 12) $-20 - 56i$ | 19) $-\frac{2}{17} + \frac{8}{17}i$ | 25) $-\frac{10}{17} + \frac{11}{17}i$ |
| 5) $-2 - 11i$ | 13) $30 + 24i$ | 20) $\frac{19}{20} - \frac{17}{20}i$ | |
| 6) $21 - 20i$ | 14) $\frac{i}{5}$ | 21) $-\frac{1}{2} + \frac{5}{2}i$ | |
| 7) $22 - 32i$ | 15) $\frac{4}{5} + \frac{3}{5}i$ | | |
| 8) $-35 + 12i$ | | | |

### Graphing Complex Numbers

| | | | |
|---|---|---|---|
| 1) $-4 - 3i$ | 2) $3 + i$ | 3) $-4 + 3i$ | 4) $4 - 2i$ |

### Rationalizing Imaginary Denominators

| | | | |
|---|---|---|---|
| 1) $-i$ | 7) $\frac{7}{3}i$ | 13) $\frac{32}{17} - \frac{8}{17}i$ | 19) $1 - 4i$ |
| 2) $-\frac{1}{5}i$ | 8) $\frac{17}{3}i$ | 14) $\frac{4}{5} - \frac{3}{5}i$ | 20) $-\frac{1}{5} - \frac{2}{5}i$ |
| 3) $\frac{-1}{13}i$ | 9) $-\frac{4x}{5y}i$ | 15) $2 + 3i$ | 21) $-\frac{4}{5} - \frac{11}{5}i$ |
| 4) $8i$ | 10) $5 + 5i$ | 16) $-\frac{23}{10} + \frac{1}{10}i$ | 22) $-\frac{44}{29} + \frac{6}{29}i$ |
| 5) $-\frac{5}{2}i$ | 11) $11 + 5i$ | 17) $-\frac{3}{5} - \frac{6}{5}i$ | 23) $\frac{41}{34} + \frac{11}{34}i$ |
| 6) $4i$ | 12) $1 - \frac{21}{4}i$ | 18) $-1 - 2i$ | 24) $-\frac{3}{29} - \frac{22}{29}i$ |

# Chapter 10 :
# Functions Operations and Quadratic

## Topics that you'll practice in this chapter:

- ✓ Evaluating Function
- ✓ Adding and Subtracting Functions
- ✓ Multiplying and Dividing Functions
- ✓ Composition of Functions
- ✓ Quadratic Equation
- ✓ Solving Quadratic Equations
- ✓ Quadratic Formula and the Discriminant
- ✓ Quadratic Inequalities
- ✓ Graphing Quadratic Functions
- ✓ Domain and Range of Radical Functions
- ✓ Solving Radical Equations

*It's fine to work on any problem, so long as it generates interesting mathematics along the way – even if you don't solve it at the end of the day." – Andrew Wiles*

# Evaluating Function

**✎ Write each of following in function notation.**

1) $h = -8x + 3$

2) $k = 2a - 14$

3) $d = 11t$

4) $y = \frac{5}{12}x - \frac{7}{12}$

5) $m = 24n - 210$

6) $c = p^2 - 5p + 10$

**✎ Evaluate each function.**

7) $f(x) = 2x - 7$, find $f(-3)$

8) $g(x) = \frac{1}{9}x + 12$, find $f(18)$

9) $h(x) = -4x + 9$, find $f(3)$

10) $f(x) = -x + 19$, find $f(-3)$

11) $f(a) = 7a - 12$, find $f(3)$

12) $h(x) = 14 - 3x$, find $f(-4)$

13) $g(n) = 6n - 10$, find $f(2)$

14) $f(x) = -11x - 4$, find $f(-1)$

15) $k(n) = -20 - 3.5n$, find $f(2)$

16) $f(x) = -0.7x + 3.3$, find $f(-7)$

17) $g(n) = \frac{11n+8}{n}$, find $g(2)$

18) $g(n) = \sqrt{3n} + 12$, find $g(3)$

19) $h(x) = x^{-2} - 7$, find $h(\frac{1}{9})$

20) $h(n) = n^{-3} + 11$, find $h(\frac{1}{4})$

21) $h(n) = n^3 - 2$, find $h(\frac{1}{2})$

22) $h(n) = n^2 - 4$, find $h(-\frac{1}{3})$

23) $h(n) = 4n^2 - 13$, find $h(-5)$

24) $h(n) = -2n^3 - 6n$, find $h(2)$

25) $g(n) = \sqrt{16n^2} - \sqrt{n}$, find $g(4)$

26) $h(a) = \frac{-14a+9}{3a}$, find $h(-b)$

27) $k(a) = 12a - 14$, find $k(a - 3)$

28) $h(x) = \frac{1}{9}x + 18$, find $h(-18x)$

29) $h(x) = 8x^2 + 16$, find $h(\frac{x}{2})$

30) $h(x) = x^4 - 20$, find $h(-2x)$

# Adding and Subtracting Functions

✎**Perform the indicated operation.**

1) $f(x) = 2x + 3$

   $g(x) = x + 7$

   Find $(f - g)(2)$

2) $g(a) = -5a - 8$

   $f(a) = -3a - 5$

   Find $(g - f)(-2)$

3) $h(t) = 4t + 3$

   $g(t) = 4t + 7$

   Find $(h - g)(t)$

4) $g(a) = -6a - 10$

   $f(a) = 3a^2 + 9$

   Find $(g - f)(x)$

5) $g(x) = \frac{5}{6}x - 23$

   $h(x) = \frac{5}{12}x + 25$

   Find $g(12) - h(12)$

6) $h(x) = \sqrt{3x} - 2$

   $g(x) = \sqrt{3x} + 5$

   Find $(h + g)(12)$

7) $f(x) = x^{-1}$

   $g(x) = x^2 + \frac{5}{x}$

   Find $(f - g)(-3)$

8) $h(n) = n^2 + 2$

   $g(n) = -4n + 6$

   Find $(h - g)(2a)$

9) $g(x) = -2x^2 - 5 - 4x$

   $f(x) = 7 + 2x$

   Find $(g - f)(3x)$

10) $g(t) = 11t - 4$

    $f(t) = -2t^2 + 5$

    Find $(g + f)(-t)$

11) $f(x) = 8x + 9$

    $g(x) = -5x^2 + 3x$

    Find $(f - g)(-x^2)$

12) $f(x) = -3x^4 - 5x$

    $g(x) = 2x^4 + 5x + 22$

    Find $(f + g)(3x^2)$

## Multiplying and Dividing Functions

✏ **Perform the indicated operation.**

1) $g(x) = -2x - 1$

$f(x) = 4x + 3$

Find $(g.f)(2)$

2) $f(x) = 5x$

$h(x) = -2x + 3$

Find $(f.h)(-2)$

3) $g(a) = 5a - 2$

$h(a) = 2a - 3$

Find $(g.h)(-3)$

4) $f(x) = 2x - 7$

$h(x) = x - 5$

Find $(\frac{f}{h})(4)$

5) $f(x) = 8a^2$

$g(x) = 3 + 2a$

Find $(\frac{f}{g})(2)$

6) $g(a) = \sqrt{4a} + 2$

$f(a) = (-a)^4 + 1$

Find $(\frac{g}{f})(1)$

7) $g(t) = t^3 + 1$

$h(t) = 5t - 2$

Find $(g.h)(-2)$

8) $g(n) = n^2 + 2n - 4$

$h(n) = -5n + 3$

Find $(g.h)(1)$

9) $g(a) = (a - 3)^2$

$f(a) = a^2 + 4$

Find $(\frac{g}{f})(3)$

10) $g(x) = -3x^2 + \frac{4}{5}x + 9$

$f(x) = x^2 - 24$

Find $(\frac{g}{f})(5)$

11) $f(x) = 2x^3 - 5x^2 + 1$

$g(x) = 3x - 1$

Find $(f.g)(x)$

12) $f(x) = 5x - 2$

$g(x) = x^3 - 2x$

Find $(f.g)(x^2)$

# Composition of Functions

✎ Using $f(x) = 2x - 5$ and $g(x) = -2x$, find:

1) $f(g(2)) =$

4) $g(f(5)) =$

2) $f(g(-1)) =$

5) $f(g(3)) =$

3) $g(f(-4)) =$

6) $g(f(0)) =$

✎ Using $f(x) = -\frac{1}{4}x + \frac{3}{4}$ and $g(x) = 2x^2$, find:

7) $g(f(-2)) =$

10) $f(f(1)) =$

8) $g(f(4)) =$

11) $g(f(-4)) =$

9) $g(g(1)) =$

12) $f(g(x)) =$

✎ Using $f(x) = -2x + 2$ and $g(x) = x + 1$, find:

13) $g(f(1)) =$

16) $f(g(-3)) =$

14) $f(f(0)) =$

17) $g(f(2)) =$

15) $f(g(-1)) =$

18) $f(g(x)) =$

✎ Using $f(x) = \sqrt{x + 9}$ and $g(x) = x - 9$, find:

19) $f(g(9)) =$

22) $f(f(7)) =$

20) $g(f(-9)) =$

23) $g(f(-5)) =$

21) $f(g(4)) =$

24) $g(g(0)) =$

# Quadratic Equation

## ✎ Multiply.

1) $(x - 4)(x + 6) = $ _____

2) $(x + 5)(x + 7) = $ _____

3) $(x - 6)(x + 8) = $ _____

4) $(x + 2)(x - 9) = $ _____

5) $(x - 7)(x - 8) = $ _____

6) $(3x + 2)(x - 3) = $ _____

7) $(4x - 3)(x + 2) = $ _____

8) $(4x - 5)(x + 1) = $ _____

9) $(7x + 1)(x - 6) = $ _____

10) $(5x + 1)(3x - 3) = $ _____

## ✎ Factor each expression.

11) $x^2 - 2x - 8 = $ _____

12) $x^2 + 8x + 15 = $ _____

13) $x^2 - 2x - 24 = $ _____

14) $x^2 - 10x + 21 = $ _____

15) $x^2 + 10x + 21 = $ _____

16) $4x^2 + 9x + 5 = $ _____

17) $5x^2 + 13x - 6 = $ _____

18) $5x^2 + 17x - 12 = $ _____

19) $2x^2 + 7x + 5 = $ _____

20) $9x^2 - 21x + 6 = $ _____

## ✎ Calculate each equation.

21) $(x + 6)(x - 3) = 0$

22) $(x + 1)(x + 8) = 0$

23) $(3x + 6)(x + 5) = 0$

24) $(2x - 2)(4x + 8) = 0$

25) $x^2 + x + 10 = 22$

26) $x^2 + 11x + 36 = 12$

27) $2x^2 + 9x + 9 = 5$

28) $x^2 + 3x - 24 = 4$

29) $5x^2 + 5x - 40 = 20$

30) $8x^2 + 8x = 48$

## Solving Quadratic Equations

✎ **Solve each equation by factoring or using the quadratic formula.**

1) $(x + 9)(x - 1) = 0$

2) $(x + 7)(x + 6) = 0$

3) $(x - 8)(x + 3) = 0$

4) $(x - 6)(x - 4) = 0$

5) $(x + 2)(x + 12) = 0$

6) $(5x + 4)(x + 7) = 0$

7) $(6x + 1)(4x + 5) = 0$

8) $(2x + 7)(x + 8) = 0$

9) $(x + 6)(3x + 15) = 0$

10) $(12x + 2)(x + 8) = 0$

11) $x^2 = 8x$

12) $x^2 - 16 = 0$

13) $3x^2 + 6 = 9x$

14) $-2x^2 - 8 = 10x$

15) $5x^2 + 40x = 45$

16) $x^2 + 10x = 24$

17) $x^2 + 6x = 16$

18) $x^2 + 9x = -18$

19) $x^2 + 13x = -36$

20) $x^2 + 3x - 15 = 5x$

21) $x^2 + 8x + 7 = -8$

22) $3x^2 - 11x = -9 + x$

23) $10x^2 + 3 = 27x - 15$

24) $7x^2 - 6x + 8 = 8$

25) $2x^2 - 12 = -3x + 2$

26) $10x^2 - 26x - 3 = -15$

27) $3x^2 + 21 = -16x + 5$

28) $x^2 + 15x - 10 = -66$

29) $3x^2 - 8x - 8 = 4 + x$

30) $2x^2 + 6x - 24 = 12$

31) $3x^2 - 33x + 54 = -18$

32) $-10x^2 - 15x - 9 = -9 - 27x^2$

# Quadratic Formula and the Discriminant

✎ **Find the value of the discriminant of each quadratic equation.**

1) $3x(x - 8) = 0$

11) $5x^2 + 2x - 3 = 0$

2) $2x^2 + 6x - 4 = 0$

12) $-3x^2 - 11x + 4 = 0$

3) $x^2 + 6x + 7 = 0$

13) $-6x^2 - 12x + 8 = 0$

4) $x^2 - x + 3 = 0$

14) $-x^2 - 9x - 12 = 0$

5) $x^2 + 4x - 3 = 0$

15) $7x^2 - 6x - 10 = 0$

6) $2x^2 + 6x - 10 = 0$

16) $-4x^2 - 2x + 8 = 0$

7) $3x^2 + 7x + 5 = 0$

17) $5x^2 + 8x - 2 = 0$

8) $x^2 - 6x - 4 = 0$

18) $6x^2 - 4x = 0$

9) $2x^2 + 8x + 3 = 0$

19) $3x^2 - 5x + 2 = 0$

10) $x^2 + 7x - 5 = 0$

20) $4x^2 + 9x + 3 = 0$

✎ **Find the discriminant of each quadratic equation then state the number of real and imaginary solutions.**

21) $-4x^2 - 16 = 16x$

25) $-11x^2 = -15x + 8$

22) $20x^2 = 20x - 5$

26) $3x^2 + 6x + 9 = 6$

23) $-11x^2 - 19x = 26$

27) $13x^2 - 5x - 12 = -26$

24) $22x^2 - 4x + 1 = 18x^2$

28) $-8x^2 - 32x - 25 = 7$

# Graphing Quadratic Functions

✍ketch the graph of each function. Identify the vertex and axis of symmetry.

1) $y = (x + 3)^2 + 2$

2) $y = (x - 3)^2 - 2$

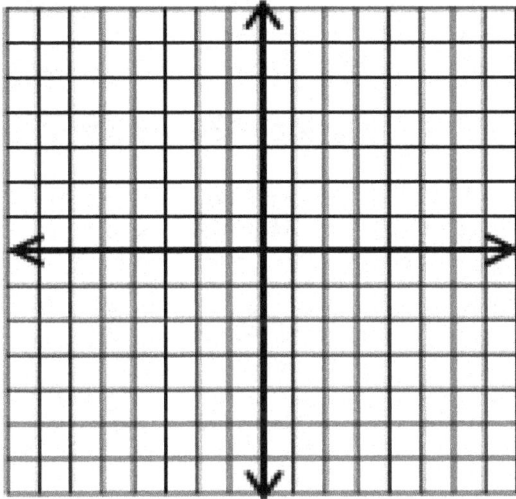
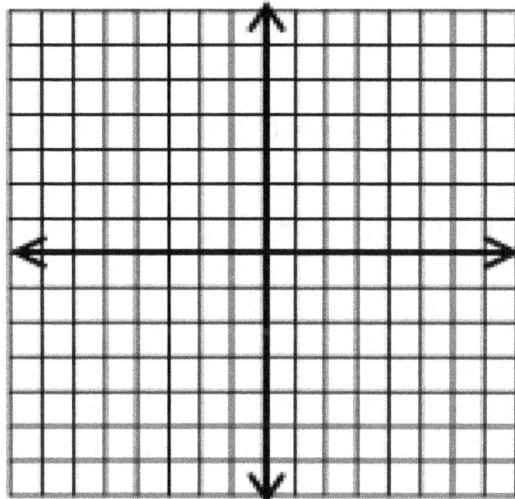

3) $y = 6 - (-x + 4)^2$

4) $y = -3x^2 - 6x + 9$

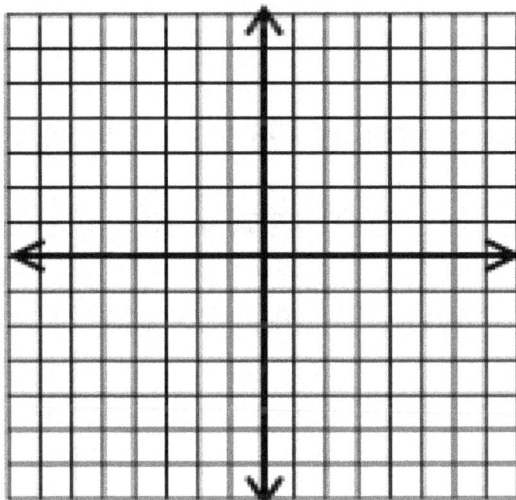
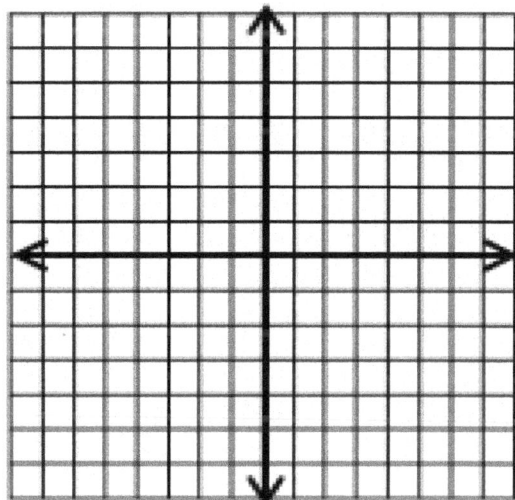

# Quadratic Inequalities

✎ Solve each quadratic inequality.

1) $x^2 - 25 < 0$

2) $-x^2 - 6x - 8 > 0$

3) $5x^2 + 15x + 30 < 0$

4) $x^2 + 8x + 16 > 0$

5) $2x^2 - 18x - 20 \geq 0$

6) $x^2 > -10x - 25$

7) $3x^2 + 2x + 16 \leq 0$

8) $x^2 - 5x - 14 \leq 0$

9) $x^2 - 6x - 7 \geq 0$

10) $2x^2 + 16x - 18 < 0$

11) $x^2 + 6x - 72 > 0$

12) $3x^2 - 3x - 36 > 0$

13) $x^2 - 15x + 64 \leq 0$

14) $2x^2 - 24x + 72 \leq 0$

15) $x^2 - 16x + 63 \geq 0$

16) $x^2 - 16x + 55 \geq 0$

17) $x^2 - 81 \leq 0$

18) $x^2 - 17x + 42 \geq 0$

19) $9x^2 + 14x + 36 \leq 0$

20) $4x^2 - 2x - 24 > 2x^2$

21) $5x^2 - 20x + 20 < 0$

22) $7x^2 - 6x \geq 6x^2 - 5$

23) $5x^2 - 15 > 4x^2 + 2x$

24) $3x^2 - 4x \geq 3x^2 - 9x + 15$

25) $8x^2 + 9x - 54 > 5x^2$

26) $10x^2 + 50x - 60 < 0$

27) $-x^2 + 15x - 57 \geq 0$

28) $-5x^2 + 25x + 30 \leq 0$

29) $5x^2 + 40x + 75 < 0$

30) $9x^2 + 20x + 180 \leq 0$

31) $3x^2 + 2x - 36 \geq -x$

32) $3x^2 + 9x + 9 \leq 6x^2 + 3x$

# Domain and Range of Radical Functions

✎ **Identify the domain and range of each function.**

1) $y = \sqrt{x + 8} - 7$

2) $y = \sqrt[3]{3x - 5} - 4$

3) $y = \sqrt{3x - 9} + 3$

4) $y = \sqrt[3]{(4x + 6)} - 2$

5) $y = 3\sqrt{4x + 20} + 6$

6) $y = \sqrt[3]{(5x - 2)} - 11$

7) $y = 4\sqrt{9x^2 + 8} + 3$

8) $y = \sqrt[3]{(7x^2 - 2)} - 6$

9) $y = 2\sqrt{2x^3 + 16} - 3$

10) $y = \sqrt[3]{(11x + 4)} - 2x$

11) $y = 3\sqrt{-2(4x + 8)} + 5$

12) $y = \sqrt[5]{(3x^2 - 12)} - 6$

13) $y = 3\sqrt{x - 5} - 2$

14) $y = \sqrt[3]{6x + 9} - 4$

✎ **Sketch the graph of each function.**

15) $y = -3\sqrt{x} + 5$

16) $y = 3\sqrt{x} - 6$

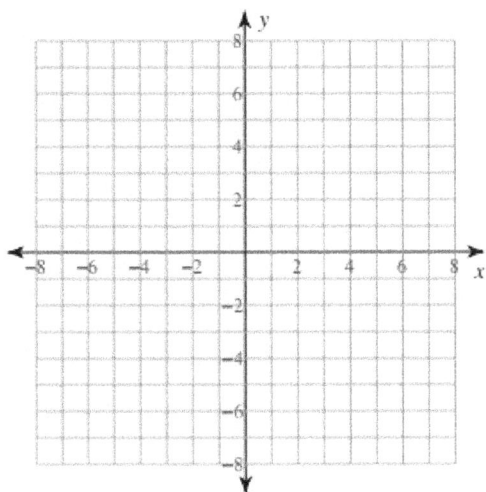

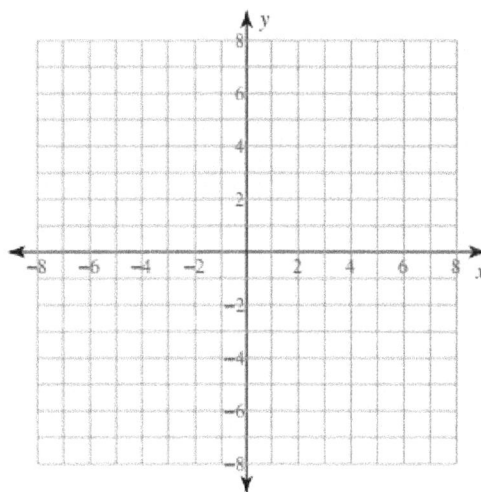

## Solving Radical Equations

✍ **Solve each equation. Remember to check for extraneous solutions.**

1) $\sqrt{a} = 9$

2) $\sqrt{v} = 6$

3) $\sqrt{r} = 4$

4) $8 = 16\sqrt{x}$

5) $\sqrt{x+3} = 18$

6) $6 = \sqrt{x-7}$

7) $4 = \sqrt{r-3}$

8) $\sqrt{x-5} = 7$

9) $12 = \sqrt{x-4}$

10) $\sqrt{m+5} = 8$

11) $7\sqrt{5a} = 35$

12) $6\sqrt{2x} = 48$

13) $2 = \sqrt{6x-32}$

14) $\sqrt{304-4x} = 4$

15) $\sqrt{r+2} - 8 = 6$

16) $-21 = -7\sqrt{r+9}$

17) $60 = 6\sqrt{5v}$

18) $x = \sqrt{40-3x}$

19) $\sqrt{90-27a} = 3a$

20) $\sqrt{-8n+88} = 4$

21) $\sqrt{15r-5} = 4r-3$

22) $\sqrt{-64+32x} = 4x$

23) $\sqrt{4x+15} = \sqrt{2x+11}$

24) $\sqrt{12v} = \sqrt{15v-21}$

25) $\sqrt{9-x} = \sqrt{x-3}$

26) $\sqrt{6m+34} = \sqrt{8m+34}$

27) $\sqrt{7r+32} = \sqrt{-8-3r}$

28) $\sqrt{4k+10} = \sqrt{2-4k}$

29) $-20\sqrt{x-13} = -40$

30) $\sqrt{90-2x} = \sqrt{\dfrac{x}{4}}$

# Answers of Worksheets

## Evaluating Function

1) $h(x) = -8x + 3$
2) $k(a) = 2a - 14$
3) $d(t) = 11t$
4) $f(x) = \frac{5}{12}x - \frac{7}{12}$
5) $m(n) = 24n - 210$
6) $c(p) = p^2 - 5p + 10$
7) $-13$
8) $14$
9) $-3$
10) $22$

11) $9$
12) $26$
13) $2$
14) $7$
15) $-27$
16) $8.2$
17) $15$
18) $15$
19) $74$
20) $75$

21) $-1\frac{7}{8}$
22) $-3\frac{8}{9}$
23) $87$
24) $-28$
25) $14$
26) $-\frac{14b+9}{3b}$
27) $12a - 50$
28) $-2x + 18$
29) $2x^2 + 16$
30) $16x^4 - 20$

## Adding and Subtracting Functions

1) $-2$
2) $1$
3) $-4$
4) $-3x^2 - 6x - 19$

5) $-43$
6) $15$
7) $-7\frac{2}{3}$
8) $4a^2 + 8a - 4$

9) $-18x^2 - 18x - 12$
10) $-2t^2 - 11t + 1$
11) $5x^4 - 5x^2 + 9$
12) $-81x^8 + 22$

## Multiplying and Dividing Functions

1) $-55$
2) $-70$
3) $153$
4) $-1$

5) $4\frac{4}{7}$
6) $2$
7) $84$

8) $2$
9) $0$
10) $-62$

11) $6x^4 - 17x^3 + 5x^2 + 3x - 1$

12) $5x^8 - 2x^6 - 10x^4 + 4x^2$

## Composition of Functions

1) $-13$
2) $-1$
3) $26$
4) $-10$
5) $-17$

6) $10$
7) $\frac{25}{8}$
8) $\frac{1}{8}$
9) $8$

10) $\frac{5}{8}$
11) $\frac{49}{8}$
12) $-\frac{1}{2}(x^2 - \frac{3}{2})$
13) $1$

14) $-2$
15) $2$
16) $6$
17) $-1$
18) $-2x$

19) 3    21) 2    23) $-7$

20) $-9$    22) $\sqrt{13}$    24) $-18$

## Quadratic Equations

1) $x^2 + 2x - 24$
2) $x^2 + 12x + 35$
3) $x^2 + 2x - 48$
4) $x^2 - 7x - 18$
5) $x^2 - 15x + 56$
6) $3x^2 - 7x - 6$
7) $4x^2 + 5x - 6$
8) $4x^2 - x - 5$
9) $7x^2 - 41x - 6$
10) $15x^2 - 12x - 3$

11) $(x - 4)(x + 2)$
12) $(x + 5)(x + 3)$
13) $(x - 6)(x + 4)$
14) $(x - 3)(x - 7)$
15) $(x + 3)(x + 7)$
16) $(4x + 5)(x + 1)$
17) $(5x - 2)(x + 3)$
18) $(5x - 3)(x + 4)$
19) $(2x + 5)(x + 1)$
20) $3(x - 2)(3x - 1)$

21) $x = -6, x = 3$
22) $x = -1, \dot{x} = -8$
23) $x = -2, x = -5$
24) $x = 1, x = -2$
25) $x = 3, x = -4$
26) $x = -3, x = -8$
27) $x = -4, x = -\frac{1}{2}$
28) $x = 4, x = -7$
29) $x = 3, x = -4$
30) $x = -3, x = 2$

## Solving quadratic equations

1) $\{-9, 1\}$
2) $\{-6, -7\}$
3) $\{8, -3\}$
4) $\{6, 4\}$
5) $\{-2, -12\}$
6) $\{-\frac{4}{5}, -7\}$
7) $\{-\frac{5}{4}, -\frac{1}{6}\}$
8) $\{-\frac{7}{2}, -8\}$

9) $\{-6, -5\}$
10) $\{-\frac{1}{6}, -8\}$
11) $\{8, 0\}$
12) $\{4, -4\}$
13) $\{2, 1\}$
14) $\{-4, -1\}$
15) $\{1, -9\}$
16) $\{2, -12\}$

17) $\{2, -8\}$
18) $\{-3, -6\}$
19) $\{-4, -9\}$
20) $\{5, -3\}$
21) $\{-5, -3\}$
22) $\{1, 3\}$
23) $\{\frac{6}{5}, \frac{3}{2}\}$
24) $\{\frac{6}{7}, 0\}$

25) $\{-\frac{7}{2}, 2\}$
26) $\{\frac{3}{5}, 2\}$
27) $\{-\frac{4}{3}, -4\}$
28) $\{-8, -7\}$
29) $\{4, -1\}$
30) $\{3, -6\}$
31) $\{3, 8\}$
32) $\{\frac{15}{17}, 0\}$

## Quadratic formula and the discriminant

1) 576
2) 68
3) 8
4) $-11$
5) 28

6) 116
7) $-11$
8) 52
9) 40
10) 69

11) 64
12) 169
13) 336
14) 33
15) 316

16) 132
17) 104
18) 16
19) 1
20) 33

21) $0,$ *one real solution*    22) $0,$ *one real solution*    23) $-783,$ *no solution*

24) $0$, *one real solution*    26) $0$, *one real solution*    28) $0$, *one real solution*

25) $-127$, *no solution*    27) $-703$, *no solution*

**Graphing quadratic functions**

1)  $(-3, 2), x = -3$

2)  $(3, -2), x = 3$

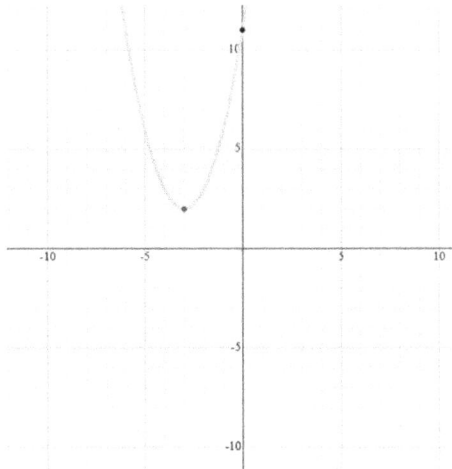    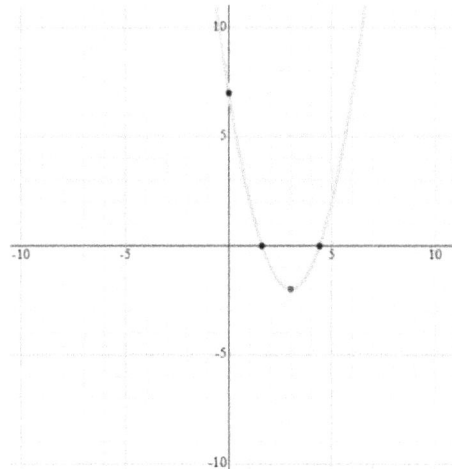

3)  $(4, 6), x = 4$

4)  $(-1, 12), x = -1$

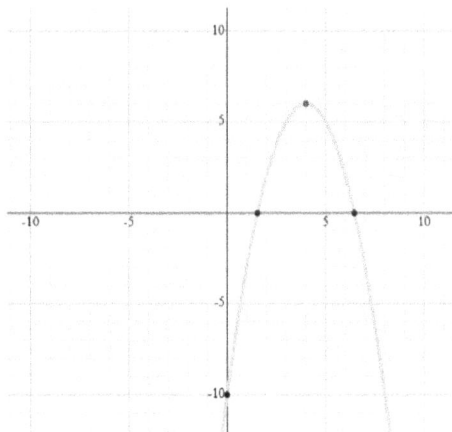    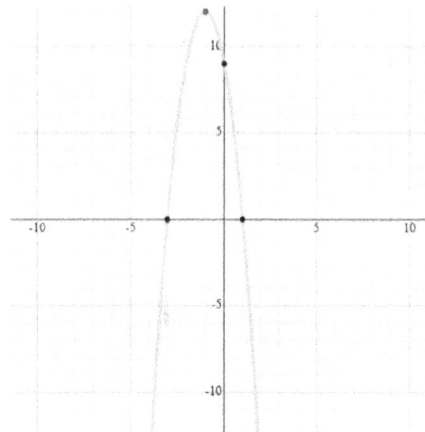

**Quadratic inequalities**

1)  $-5 < x < 5$

7)  no solution

13) no solution

2)  $-4 < x < -2$

8)  $-2 \leq x \leq 7$

14) $x = 6$

3)  no solution

9)  $x \leq -1 \, or \, x \geq 7$

15) $x \leq 7 or \ x \geq 9$

4)  $x < -4 \, or \, x > -4$

10) $-9 < x < 1$

16) $x \leq 5 or \ x \geq 11$

5)  $x \leq -1 \, or \, x \geq 10$

11) $x < -12 \, or \, x > 6$

17) $-9 \leq x \leq 9$

6)  $x < -5 \, or \, x > -5$

12) $-3 < x < 4$

18) $x \leq 3 \, or \, x \geq 14$

19) no solution

20) $x < -3$ or $x > 4$

21) no solution

22) $x \leq 1$ or $x \geq 5$

23) $x < -3$ or $x > 5$

24) $x \geq 3$

25) $x < -6$ or $x > 3$

26) $-6 < x < 1$

27) no solution

28) $x \leq -1$ or $x \geq 6$

29) $-5 < x < -3$

30) no solution

31) $x \leq -4$ or $x \geq 3$

32) $x \leq -1$ or $x \geq 3$

## Domain and range of radical functions

1) domain: $x \geq -8$

   range: $y \geq -7$

2) domain: {all real numbers}

   range: {all real numbers}

3) domain: $x \geq 3$

   range: $y \geq 3$

4) domain: {all real numbers}

   range: {all real numbers}

5) domain: $x \geq -5$

   range: $y \geq 6$

6) domain: {all real numbers}

   range: {all real numbers}

7) domain: {all real numbers}

   range: $y \geq 8\sqrt{2} + 3$

8) domain: {all real numbers}

   range: {all real numbers}

9) domain: $x \geq -2$

   range: $y \geq -3$

10) domain: {all real numbers}

    range: {all real numbers}

11) domain: $x \leq -2$

    range: $y \geq 5$

12) domain: {all real numbers}

    range: {all real numbers}

13) domain: $x \geq 5$

    range: $y \geq -2$

14) domain: {all real numbers}

    range: {all real numbers}

15)

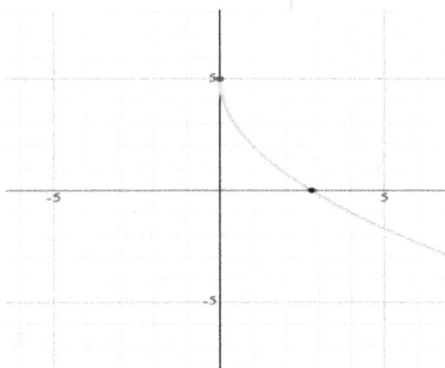

16)

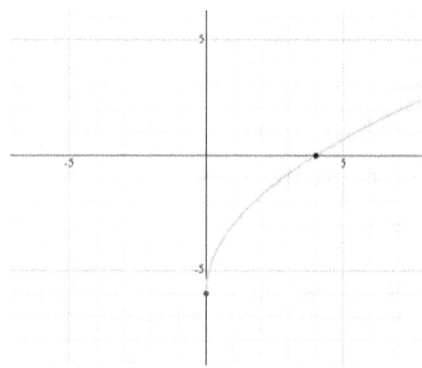

**Solving radical equations**

1) $\{81\}$

2) $\{36\}$

3) $\{16\}$

4) $\{\frac{1}{4}\}$

5) $\{321\}$

6) $\{43\}$

7) $\{19\}$

8) $\{54\}$

9) $\{148\}$

10) $\{59\}$

11) $\{5\}$

12) $\{32\}$

13) $\{6\}$

14) $\{72\}$

15) $\{194\}$

16) $\{0\}$

17) $\{20\}$

18) $\{5\}$

19) $\{2\}$

20) $\{9\}$

21) $\{2\}$

22) no solution

23) $\{-2\}$

24) $\{7\}$

25) $\{6\}$

26) $\{0\}$

27) $\{-4\}$

28) $\{-1\}$

29) $\{17\}$

30) $\{40\}$

# Chapter 11 :

# Sequences and Series

## Topics that you'll practice in this chapter:

✓ Arithmetic Sequences

✓ Geometric Sequences

✓ Comparing Arithmetic and Geometric Sequences

✓ Finite Geometric Series

✓ Infinite Geometric Series

*"The fact that people will be full of greed, fear, or folly is predictable. The sequence is not predictable." -Warren Buffett*

# Arithmetic Sequences

✎ **Find the next three terms of each arithmetic sequence.**

1) $32, 26, 20, 14, 8, \ldots$

2) $-56, -44, -32, -20, \ldots$

3) $17, 26, 35, 44, 53, \ldots$

4) $5, 11, 17, 23, 29, \ldots$

✎ **Given the first term and the common difference of an arithmetic sequence find the first five terms and the explicit formula.**

5) $a_1 = 20, d = 3$

7) $a_1 = 32, d = 6$

6) $a_1 = -11, d = -5$

8) $a_1 = 240, d = -80$

✎ **Given a term in an arithmetic sequence and the common difference find the first five terms and the explicit formula.**

9) $a_{20} = -500, d = -50$

11) $a_{51} = -88.2, d = -5.2$

10) $a_{24} = 98, d = 7$

12) $a_{68} = -980, d = -27$

✎ **Given a term in an arithmetic sequence and the common difference find the recursive formula and the three terms in the sequence after the last one given.**

13) $a_{21} = -187, d = -9$

15) $a_{31} = 58.2, d = 1.8$

14) $a_{12} = 63.5, d = 5.2$

16) $a_{42} = 6.8, d = 0.4$

## Geometric Sequences

✍ **Determine if the sequence is geometric. If it is, find the common ratio.**

1) $1, -7, 49, -343, \dots$

3) $8, 24, 48, 240, \dots$

2) $-3, -12, -48, -192, \dots$

4) $-5, -10, -20, -40, \dots$

✍ **Given the first term and the common ratio of a geometric sequence find the first five terms and the explicit formula.**

5) $a_1 = 0.4, r = -3$

6) $a_1 = 0.2, r = 4$

✍ **Given the recursive formula for a geometric sequence find the common ratio, the first five terms, and the explicit formula.**

7) $a_n = a_{n-1} \times 4, a_1 = 2$

9) $a_n = a_{n-1} \cdot 5, a_1 = 0.2$

8) $a_n = a_{n-1} \cdot (-2), a_1 = -4$

10) $a_n = a_{n-1} \cdot 3, a_1 = -3$

✍ **Given two terms in a geometric sequence find the 6th term and the recursive formula.**

11) $a_3 = 576$ and $a_5 = 36$

12) $a_2 = -0.4$ and $a_4 = -1.6$

# Comparing Arithmetic and Geometric Sequences

✎ **For each sequence, state if it is arithmetic, geometric, or neither.**

1) $6, 11, 16, 21, \ldots$

2) $2, 5, 8, 11, \ldots$

3) $3, 10, 20, 27, \ldots$

4) $1, 11, 22, 33, 44, \ldots$

5) $4, 8, 10, 24, 96, \ldots$

6) $2, 10, 20, 50, 200, \ldots$

7) $0.2, 1, 5, 25, 125, \ldots$

8) $4, 12, 36, 108, \ldots$

9) $-27, -34, -41, -48, -55, \ldots$

10) $-3, 15, -75, 375, -1875, \ldots$

11) $10, 25, 50, 65, 80, \ldots$

12) $5, 15, 150, 250, 350 \ldots$

13) $-35, -20, -5, 10, 25, \ldots$

14) $a_n = 4 . 8^{n-1}$

15) $a_n = 3 . 6^{n-1}$

16) $a_n = 8 - 4n$

17) $a_n = -210 + 310n$

18) $a_n = 53 + 53n$

19) $a_n = -10 . (-5)^{n-1}$

20) $a_n = 49 + 63n$

21) $a_n = (3n)^4$

22) $a_n = 40 + 8n$

23) $a_n = -(13)^{n-1}$

24) $a_n = -10 . (0.2)^{n-1}$

25) $a_n = \frac{3n+2}{4^n}$

26) $a_n = \frac{47+13n}{7^n}$

27) $a_n = \frac{12-12n}{12^n}$

28) $a_n = \frac{32-a_{n-1}}{2n}$

29) $a_n = \frac{4}{15} - \frac{2}{7}n$

## Finite Geometric Series

✎ **Evaluate the related series of each sequence.**

1) $-2, 4, -8, 16$

2) $-1, 6, -36, 216, -1,296$

3) $-2, 8, -32, 128, -512$

4) $4, 8, 16, 32, 64$

5) $-4, -12, -36, -108$

6) $5, -15, 45, -135, 405$

✎ **Evaluate each geometric series described.**

7) $1 + 5 + 25 + 125 \ldots, n = 5$ _____

8) $1 - 6 + 36 - 216 \ldots, n = 6$ _____

9) $-2 - 6 - 18 - 54 \ldots, n = 8$ _____

10) $0.2 - 1 + 5 - 25 \ldots, n = 6$ _____

11) $0.6 - 3.6 + 21.6 - 129.6 \ldots, n = 5$ _____

12) $-1 - 4 - 16 - 64 \ldots, n = 6$ _____

13) $a_1 = -2, r = 6, n = 5$ _____

14) $a_1 = 1, r = 9, n = 6$ _____

15) $\sum_{n=1}^{6} 4 \cdot (-3)^{n-1}$ _____

16) $\sum_{n=1}^{7} 2 \cdot (-5)^{n-1}$ _____

17) $\sum_{n=1}^{10} 0.1 \cdot (2)^{n-1}$ _____

18) $\sum_{m=1}^{6} (-4)^{m-1}$ _____

19) $\sum_{m=1}^{6} 5 \times (3)^{m-1}$ _____

20) $\sum_{k=1}^{5} 7 \times (6)^{k-1}$ _____

## Infinite Geometric Series

✍ **Determine if each geometric series converges or diverges.**

1) $a_1 = -1.8$, $r = 6$

2) $a_1 = 10.8$, $r = 0.3$

3) $a_1 = -2$, $r = 6.1$

4) $a_1 = 5$, $r = 0.24$

5) $a_1 = 1.2$, $r = 8$

6) $-1, 6, -36, 216, \ldots$

7) $6, -1, \frac{1}{6}, -\frac{1}{36}, \frac{1}{216}, \ldots$

8) $512 + 64 + 8 + 1 \ldots$

9) $-5 + \frac{15}{7} - \frac{45}{49} + \frac{135}{343} \cdots$

10) $\frac{400}{459} - \frac{200}{153} + \frac{100}{51} - \frac{50}{17} \cdots$

✍ **Evaluate each infinite geometric series described.**

11) $a_1 = 2$, $r = -\frac{1}{4}$

12) $a_1 = 36$, $r = -\frac{1}{6}$

13) $a_1 = 9$, $r = \frac{1}{3}$

14) $a_1 = 12$, $r = \frac{1}{7}$

15) $1 + 0.2 + 0.04 + 0.008 + \cdots$

16) $64 - 16 + 4 - 1 \ldots,$

17) $1 - 0.3 + 0.09 - 0.027 \ldots,$

18) $-5 + \frac{15}{7} - \frac{45}{49} + \frac{135}{343} \ldots,$

19) $\sum_{k=1}^{\infty} 7^{k-1}$

20) $\sum_{i=1}^{\infty} \left(\frac{2}{5}\right)^{i-1}$

21) $\sum_{k=1}^{\infty} \left(-\frac{2}{9}\right)^{k-1}$

22) $\sum_{n=1}^{\infty} 6\left(\frac{1}{3}\right)^{n-1}$

# Answers of Worksheets

## Arithmetic Sequences

1) $2, -4, -10$

2) $-8, 4, 16$

3) $62, 71, 80$

4) $35, 41, 47$

5) First Five Terms: $20, 23, 26, 29, 32$, Explicit: $a_n = 20 + 3(n-1)$

6) First Five Terms: $-11, -16, -21, -26, -31$, Explicit: $a_n = -11 - 5(n-1)$

7) First Five Terms: $32, 38, 44, 50, 56$, Explicit: $a_n = 32 + 6(n-1)$

8) First Five Terms: $240, 160, 80, 0, -80$, Explicit: $a_n = 240 - 80(n-1)$

9) First Five Terms: $450, 400, 350, 300, 250$, Explicit: $a_n = 450 - 50(n-1)$

10) First Five Terms: $-63, -56, -49, -42, -35$, Explicit: $a_n = -63 + 7(n-1)$

11) First Five Terms: $171.8, 166.6, 161.4, 156.2, 151$, Explicit: $a_n = 171.8 - 5.2(n-1)$

12) First Five Terms: $829, 802, 775, 748, 721$, Explicit: $a_n = 829 - 27(n-1)$

13) Next 3 terms: $-196, -205, -214$, Recursive: $a_n = a_{n-1} - 9, a_1 = -7$

14) Next 3 terms: $68.7, 73.9, 79.1, 84.3$ Recursive: $a_n = a_{n-1} + 5.2, \ a_1 = 6.3$

15) Next 3 terms: $60, 61.8, 63.6$, Recursive: $a_n = a_{n-1} + 1.8, a_1 = 4.2$

16) Next 3 terms: $7.2, 7.6, 8$, Recursive: $a_n = a_{n-1} + 0.4, a_1 = -9.6$

## Geometric Sequences

1) $r = -7$

2) $r = 4$

3) not geometric

4) $r = 2$

5) First Five Terms: $0.4, -1.2, 3.6, -10.8, 32.4$

   Explicit: $a_n = 0.4 \times (-3)^{n-1}$

6) First Five Terms: $0.2, 0.8, 3.2, 12.8, 51.2$

   Explicit: $a_n = 0.2 \times (4)^{n-1}$

7) Common Ratio: $r = 4$

   First Five Terms: $2, 8, 32, 128, 512$

   Explicit: $a_n = 2 . (4)^{n-1}$

8) Common Ratio: $r = -2$

First Five Terms: $-4, 8, -16, 32, -64$

Explicit: $a_n = -4 \cdot (-2)^{n-1}$

9) Common Ratio: $r = 5$

First Five Terms: $0.2, 1, 5, 25, 125, 625$

Explicit: $a_n = 0.2 \cdot (5)^{n-1}$

10) Common Ratio: $r = 3$

First Five Terms: $-3, -9, -27, -81, -243$

Explicit: $a_n = -3 \cdot (3)^{n-1}$

11) $a_6 = -9$ , Recursive: $a_n = a_{n-1} \cdot (\frac{-1}{4})$, $a_1 = 9{,}216$

12) $a_6 = -6.4$, Recursive: $a_n = a_{n-1} \cdot (-2)$, $a_1 = 0.2$

**Comparing Arithmetic and Geometric Sequences**

| | | |
|---|---|---|
| 1) Arithmetic | 11) Neither | 21) Neither |
| 2) Arithmetic | 12) Neither | 22) Arithmetic |
| 3) Neither | 13) Arithmetic | 23) Geometric |
| 4) Neither | 14) Geometric | 24) Geometric |
| 5) Neither | 15) Geometric | 25) Neither |
| 6) Neither | 16) Arithmetic | 26) Neither |
| 7) Geometric | 17) Arithmetic | 27) Neither |
| 8) Geometric | 18) Arithmetic | 28) Neither |
| 9) Arithmetic | 19) Geometric | 29) Arithmetic |
| 10) Geometric | 20) Arithmetic | |

**Finite Geometric**

| | | |
|---|---|---|
| 1) 10 | 8) $-6{,}665$ | 15) $-728$ |
| 2) $-1{,}111$ | 9) $-6{,}560$ | 16) $26{,}042$ |
| 3) $-410$ | 10) $-520.8$ | 17) $102.3$ |
| 4) 124 | 11) $666.6$ | 18) $-819$ |
| 5) $-160$ | 12) $-1{,}365$ | 19) $1{,}820$ |
| 6) 305 | 13) $-3{,}110$ | 20) $10{,}885$ |
| 7) 781 | 14) $66{,}430$ | |

**Infinite Geometric**

1) Diverges

2) Converges

3) Diverges

4) Converges

5) Diverges

6) Diverges

7) Converges

8) Converges

9) Converges

10) Converges

11) $\frac{8}{5}$

12) $\frac{216}{7}$

13) $\frac{27}{2}$

14) 14

15) $\frac{5}{4}$

16) $\frac{256}{5}$

17) $\frac{3}{4}$

18) $-\frac{7}{2}$

19) Infinite

20) $\frac{5}{3}$

21) $\frac{9}{11}$

22) 9

# Chapter 12 :

# Logarithms

**Topics that you'll practice in this chapter:**

- ✓ Rewriting Logarithms
- ✓ Evaluating Logarithms
- ✓ Properties of Logarithms
- ✓ Natural Logarithms
- ✓ Exponential Equations Requiring Logarithms
- ✓ Solving Logarithmic Equations

*Mathematics is an art of human understanding.*

*— William Thurston*

# Rewriting Logarithms

✎ **Rewrite each equation in exponential form.**

1) $\log_3 27 = 3$

2) $\log_2 128 = 7$

3) $\log_6 1{,}296 = 4$

4) $\log_5 625 = 4$

5) $\log_{11} 121 = 2$

6) $\log_{12} 1{,}728 = 3$

7) $\log_9 729 = 3$

8) $\log_3 729 = 6$

9) $\log_{10} 10{,}000 = 4$

10) $\log_7 343 = 3$

11) $\log_4 1{,}024 = 5$

12) $\log_{12} 144 = 2$

13) $\log_{13} 2{,}197 = 3$

14) $\log_{25} 5 = \frac{1}{2}$

15) $\log_{81} 3 = \frac{1}{4}$

16) $\log_{3{,}125} 5 = \frac{1}{5}$

17) $\log_{1{,}000} 10 = \frac{1}{3}$

18) $\log_5 \frac{1}{125} = -3$

19) $\log_4 \frac{1}{16} = -2$

20) $\log_a \frac{7}{4} = b$

✎ **Rewrite each exponential equation in logarithmic form.**

21) $2^5 = 32$

22) $4^3 = 64$

23) $5^4 = 625$

24) $11^3 = 1{,}331$

25) $3^5 = 243$

26) $6^4 = 1{,}296$

27) $7^4 = 2{,}401$

28) $9^3 = 729$

29) $4^{-5} = \frac{1}{1{,}024}$

30) $3^{-8} = \frac{1}{6{,}561}$

31) $11^{-2} = \frac{1}{121}$

32) $12^{-3} = \frac{1}{1{,}728}$

33) $4^{-5} = \frac{1}{1{,}024}$

34) $10^{-5} = \frac{1}{100{,}000}$

# Evaluating Logarithms

✎ **Evaluate each logarithm.**

1) $\log_3 729 =$

2) $\log_2 256 =$

3) $\log_3 243 =$

4) $\log_4 64 =$

5) $\log_8 64 =$

6) $\log_{11} 121 =$

7) $\log_{10} 10{,}000 =$

8) $\log_5 \frac{1}{25} =$

9) $\log_4 \frac{1}{256} =$

10) $\log_2 \frac{1}{32} =$

11) $\log_6 \frac{1}{36} =$

12) $\log_9 \frac{1}{81} =$

13) $\log_{12} \frac{1}{144} =$

14) $\log_{1{,}000} \frac{1}{10} =$

15) $\log_{243} 3 =$

16) $\log_4 \frac{1}{16} =$

17) $\log_8 \frac{1}{512} =$

18) $\log_3 \frac{1}{81} =$

✎ **Circle the points which are on the graph of the given logarithmic functions.**

19) $y = 4\log_4(3x - 2) + 1$     $(3, 4)$,     $(2, 5)$,     $(7, 4)$

20) $y = 5\log_6(12x) - 7$     $(2, -2)$,     $(\frac{1}{3}, 12)$,     $(\frac{1}{2}, -2)$

21) $y = -2\log_3 9(x - 5) + 5$     $(6, -3)$,     $(8, -1)$,     $(1, 6)$

22) $y = \frac{1}{4}\log_6(6x) + \frac{1}{2}$     $(6, 1)$,     $(6, \frac{1}{4})$,     $(4, \frac{1}{4})$

23) $y = -2\log_8 8(x + 4) + 9$     $(-4, 0)$,     $(0, 9)$,     $(-2, 6\frac{1}{3})$

24) $y = -\log_5(x + 15) - 6$     $(10, -\frac{1}{5})$,     $(10, -8)$,     $(11, -\frac{2}{5})$

25) $y = -3\log_2(2x + 6) + 7$     $(5, -5)$,     $(-5, -5)$,     $(-2, -2)$

# Properties of Logarithms

✎ **Expand each logarithm.**

1) $\log (11 \times 4) =$

2) $\log (13 \times 5) =$

3) $\log (4 \times 12) =$

4) $\log \left(\frac{2}{7}\right) =$

5) $\log \left(\frac{4}{9}\right) =$

6) $\log \left(\frac{5}{8}\right)^3 =$

7) $\log (6 \times 5^4) =$

8) $\log \left(\frac{14}{3}\right)^5 =$

9) $\log \left(\frac{3^4}{8}\right) =$

10) $\log (x \times y)^8 =$

11) $\log (x^6 \times y^{12} \times z^2) =$

12) $\log \left(\frac{u^8}{v^3}\right) =$

13) $\log \left(\frac{x}{y^7}\right) =$

✎ **Condense each expression to a single logarithm.**

14) $\log 8 - \log 13 =$

15) $\log 6 + \log 11 =$

16) $4 \log 2 - 7 \log 5 =$

17) $10 \log 4 - 3 \log 7 =$

18) $3 \log 9 - \log 17 =$

19) $11 \log 6 - 9 \log 4 =$

20) $\log 15 - 6 \log 7 =$

21) $6 \log 8 + 4 \log 10 =$

22) $12 \log 5 + 14 \log 9 =$

23) $17 \log_8 a + 6 \log_8 b =$

24) $2 \log_9 x - 3 \log_9 y =$

25) $\log_{11} u - 16 \log_{11} v =$

26) $8 \log_{15} u + 9 \log_{15} v =$

27) $32 \log_6 u - 25 \log_6 v =$

# Natural Logarithms

✍ **Solve each equation for** x.

1) $e^x = 9$

2) $e^x = 36$

3) $e^x = 49$

4) $\ln x = 3$

5) $\ln (\ln x) = 7$

6) $e^x = 4$

7) $\ln(5x + 2) = 1$

8) $\ln(7x + 4) = 3$

9) $\ln(9x + 5) = 4$

10) $\ln x = \frac{1}{9}$

11) $\ln 11x = e^5$

12) $\ln x = \ln 6 + \ln 7$

13) $\ln x = 4\ln 3 + \ln 2$

✍ **Evaluate without using a calculator.**

14) $11\ln e =$

15) $\ln e^{10} =$

16) $4 \ln e =$

17) $\ln e^{21} =$

18) $32\ln e =$

19) $4\ln e^5 =$

20) $e^{\ln 22} =$

21) $e^{3\ln 3} =$

22) $e^{3\ln 5} =$

23) $\ln \sqrt[11]{e} =$

✍ **Reduce the following expressions to simplest form.**

24) $e^{-4\ln 9 + 4\ln 3} =$

25) $e^{-3\ln \left(\frac{5}{4e}\right)} =$

26) $2 \ln(e^4) =$

27) $\ln\left(\frac{1}{e}\right)^4 =$

28) $e^{\ln 9 + 3\ln 3} =$

29) $e^{\ln \left(\frac{13}{e}\right)} =$

30) $8 \ln(1^{-3e}) =$

31) $2 \ln\left(\frac{1}{e}\right)^{-3} =$

32) $6\ln \left(\frac{\sqrt[3]{e}}{3e}\right) =$

33) $e^{-4\ln e + 2\ln 5} =$

34) $e^{\ln \frac{4}{e}} =$

35) $11 \ln(e^e) =$

# Exponential Equations and Logarithms

✍ **Solve each equation for the unknown variable.**

1) $3^{4n} = 243$

2) $5^{3r} = 625$

3) $6^{2n-1} = 216$

4) $16^{2r+3} = 4$

5) $169^{2x} = 13$

6) $7^{-3v-3} = 49$

7) $2^{4n} = 128$

8) $11^{n-1} = 1,331$

9) $\frac{9^{3a}}{3^{2a}} = 729$

10) $13^5 \times 13^{-4v} = 169$

11) $4^{3n} = \frac{1}{64}$

12) $\left(\frac{1}{11}\right)^{2n} = 121$

13) $2,187^{3x} = 3$

14) $13^{5-7x} = 13^{-2x}$

15) $11^{-3x} = 11^{2x-7}$

16) $3^{5n} = 243$

17) $17^{5x+3} = 17^{6x}$

18) $15^{3n} = 225$

19) $4^{-3k} = 512$

20) $8^{-4r} = 8^{-5r+2}$

21) $8^{2x+3} = 8^{5x}$

22) $10^{3x-2} = 100,000$

23) $16 \times 64^{-v} = 128$

24) $\frac{128}{2^{-3m}} = 2^{4m+5}$

25) $14^{-5n} \times 14^{2n+3} = 14^{-2n}$

26) $\left(\frac{1}{9}\right)^{4n+3} \times \left(\frac{1}{9}\right)^{-3n-8} = \left(\frac{1}{9}\right)^{-4n}$

✍ **Solve each problem. (Round to the nearest whole number)**

27) A substance decays 16% each day. After 8 days, there are 6 milligrams of the substance remaining. How many milligrams were there initially? _____

28) A culture of bacteria grows continuously. The culture doubles every 4 hours. If the initial number of bacteria is 20, how many bacteria will there be in 13 hours? _____

29) Bob plans to invest $11,200 at an annual rate of 3.5%. How much will Bob have in the account after three years if the balance is compounded quarterly? _____

30) Suppose you plan to invest $8,000 at an annual rate of 5%. How much will you have in the account after 6 years if the balance is compounded monthly? _____

## Solving Logarithmic Equations

✍ **Find the value of the variables in each equation.**

1) $2\log(x) + 5 = 9$

2) $\log_4 4x + 3 = 5$

3) $-\log_8(8x) + 2 = 3$

4) $\log 2x - \log 4 = 1$

5) $\log 5x + \log 25 = 1$

6) $\log 4 - \log x = 3$

7) $\log 4x + \log 2 = \log 16$

8) $-6\log_3(5x - 1) = -36$

9) $\log 4x = \log(8x - 1)$

10) $\log(4k - 6) = \log(k - 3)$

11) $\log(5p + 2) = \log(p + 4)$

12) $-30 + \log_4(3n + 2) = -30$

13) $\log_4(4x - 4) = \log_4(x^2)$

14) $\log_8(k^2 + 15) = \log_8(-6k - 3)$

15) $\log(16 + 6b) = \log(10b^2 + 12b)$

16) $\log_6(2x + 5) - \log_6 x = \log_6 9$

17) $\log_5 5 + \log_5(x^2 + 1) = \log_5 25$

18) $\log_6(x + 3) + \log_6(x + 1) =$

$\quad \log_6 8$

✍ **Find the value of $x$ in each natural logarithm equation.**

19) $\ln 8 - \ln(4x + 8) = 4$

20) $\ln(x + 5) - \ln(x + 2) = \ln 10$

21) $\ln e^6 - \ln(x - 1) = 3$

22) $\ln(2x - 8) + \ln(x - 4) = \ln 8$

23) $\ln 5x - \ln(x + 4) = \ln 2$

24) $\ln(8x - 4) - \ln(x - 2) = \ln 25$

25) $\ln(3x + 2) - 4\ln 2 = 5$

26) $\ln(2x - 5) + \ln(x - 3) = \ln 6$

27) $\ln(x - 1) + \ln(4x - 7) = \ln(7)$

28) $2\ln 3x - \ln(x + 10) = \ln 2x$

29) $\ln x^4 + \ln x^8 = 4\ln(2x)$

30) $\ln x^{10} - \ln(x^2 + 10) = 10\ln 2x$

31) $8\ln(x - 2) = 4\ln(x^2 - 4x + 4)$

32) $\ln(x^4 + 10) = \ln(x^2 + 9)$

33) $2\ln x - 2\ln(x + 8) = \ln(x^2)$

34) $\ln(2x + 1) - \ln(4x + 1) = \ln 4$

35) $\ln 16 + 2\ln(x - 2) = \ln 4$

36) $\ln e^2 + \ln(5x - 6) = \ln(5) + 3$

# Answers of Worksheets

### Rewriting Logarithms

1) $3^3 = 27$

2) $2^7 = 128$

3) $6^4 = 1,296$

4) $5^4 = 625$

5) $11^2 = 121$

6) $12^3 = 1,728$

7) $9^3 = 729$

8) $3^6 = 729$

9) $10^4 = 10,000$

10) $7^3 = 343$

11) $4^5 = 1,024$

12) $12^2 = 144$

13) $13^3 = 2,197$

14) $25^{\frac{1}{2}} = 5$

15) $81^{\frac{1}{4}} = 3$

16) $3,125^{\frac{1}{5}} = 5$

17) $1,000^{\frac{1}{3}} = 10$

18) $5^{-3} = \frac{1}{125}$

19) $4^{-2} = \frac{1}{16}$

20) $a^b = \frac{7}{4}$

21) $\log_2 32 = 5$

22) $\log_4 64 = 3$

23) $\log_5 625 = 4$

24) $\log_{11} 1,331 = 3$

25) $\log_3 243 = 5$

26) $\log_6 1,296 = 4$

27) $\log_7 2,401 = 4$

28) $\log_9 729 = 3$

29) $\log_4 \frac{1}{1,024} = -5$

30) $\log_3 \frac{1}{6,561} = -8$

31) $\log_{11} \frac{1}{121} = -2$

32) $\log_{12} \frac{1}{1,728} = -3$

33) $\log_4 \frac{1}{1,024} = -5$

34) $\log_{10} \frac{1}{100,000} = -5$

### Evaluating Logarithms

1) 6

2) 8

3) 5

4) 3

5) 2

6) 2

7) 4

8) $-2$

9) $-4$

10) $-5$

11) $-2$

12) $-2$

13) $-2$

14) $-\frac{1}{3}$

15) $\frac{1}{5}$

16) $-2$

17) $-3$

18) $-4$

19) $(2, 5)$

20) $\left(\frac{1}{2}, -2\right)$

21) $(8, -1)$

22) $(6, 1)$

23) $\left(-2, 6\frac{1}{3}\right)$

24) $(10, -8)$

25) $(5, -5)$

### Properties of Logarithms

1) $\log 11 + \log 4$

2) $\log 13 + \log 5$

3) $\log 4 + \log 12$

4) $\log 2 - \log 7$

5) $\log 4 - \log 9$

6) $3 \log 5 - 3 \log 8$

7) $\log 6 + 4 \log 5$

8) $5\log 14 - 5 \log 3$

# PSAT Subject Test – Mathematics

9) $4\log 3 - \log 8$

10) $8\log x + 8\log y$

11) $6\log x + 12\log y + 2\log z$

12) $8\log u - 3\log v$

13) $\log x - 7\log y$

14) $\log \frac{8}{13}$

15) $\log(6 \times 11)$

16) $\log \frac{2^4}{5^7}$

17) $\log \frac{4^{10}}{7^3}$

18) $\log \frac{9^3}{17}$

19) $\log \frac{6^{11}}{4^9}$

20) $\log \frac{15}{7^6}$

21) $\log(8^6 \times 10^4)$

22) $\log(5^{12} \times 9^{14})$

23) $\log_8(a^{17}b^6)$

24) $\log_9 \frac{x^2}{y^3}$

25) $\log_{11} \frac{u}{v^{16}}$

26) $\log_{15}(u^8 \times v^9)$

27) $\log_6 \frac{u^{32}}{v^{25}}$

## Natural Logarithms

1) $x = \ln 9$

2) $x = \ln 36, x = 2\ln(6)$

3) $x = \ln 49, x = 2\ln(7)$

4) $x = e^3$

5) $x = e^{e^7}$

6) $x = \ln 4$

7) $x = \frac{e-2}{5}$

8) $x = \frac{e^3-4}{7}$

9) $x = \frac{e^4-5}{9}$

10) $x = \sqrt[9]{e}$

11) $x = \frac{e\,e^5}{11}$

12) $x = 42$

13) $x = 162$

14) 11

15) 10

16) 4

17) 21

18) 32

19) 20

20) 22

21) 27

22) 125

23) $\frac{1}{11}$

24) $\frac{1}{81}$

25) $\frac{64e^3}{125}$

26) 8

27) $-4$

28) 243

29) $\frac{13}{e}$

30) 0

31) 6

32) $\ln\left(\frac{1}{3^6e^4}\right) = -10.6$

33) $25e^{-4} = \frac{25}{e^4}$

34) $\frac{4}{e}$

35) $11e$

## Exponential Equations and Logarithms

1) $\frac{5}{4}$

2) $\frac{4}{3}$

3) 2

4) $-\frac{5}{4}$

5) $\frac{1}{4}$

6) $-\frac{5}{3}$

7) $\frac{7}{4}$

8) 4

9) $\frac{3}{2}$

10) $\frac{3}{4}$

11) $-1$

12) $-1$

13) $\frac{1}{21}$

14) $1$

15) $\frac{7}{5}$

16) $1$

17) $3$

18) $\frac{2}{3}$

19) $-\frac{3}{2}$

20) $2$

21) $1$

22) $\frac{7}{3}$

23) $-\frac{1}{2}$

24) $2$

25) $3$

26) $1$

27) $24.2$

28) $190.27$

29) $\$12,432.4$

30) $\$10,792.14$

**Solving Logarithmic Equations**

1) $\{100\}$

2) $\{4\}$

3) $\{\frac{1}{64}\}$

4) $\{20\}$

5) $\{\frac{2}{25}\}$

6) $\{\frac{1}{250}\}$

7) $\{2\}$

8) $\{146\}$

9) $\{\frac{1}{4}\}$

10) No Solution

11) $\{\frac{1}{2}\}$

12) $\{-\frac{1}{3}\}$

13) $\{2\}$

14) No Solution

15) $\{1, -\frac{8}{5}\}$

16) $\{\frac{5}{7}\}$

17) $\{2, -2\}$

18) $\{1\}$

19) $x = \frac{8-8e^4}{4e^4}$

20) $\{-\frac{5}{3}\}$

21) $e^3 + 1$

22) $\{6\}$

23) $\{\frac{8}{3}\}$

24) $\{\frac{46}{17}\}$

25) $x = \frac{16e^5-2}{3}$

26) $x = \frac{9}{2}$

27) $x = \frac{11}{4}$

28) $x = \frac{20}{7}$

29) $e^{\frac{\ln(2)}{2}}$

30) No Solution

31) $x > 2$

32) No Solution

33) No Solution

34) $x = -\frac{3}{14}$

35) $x = \frac{5}{2}$

36) $x = \frac{5e+6}{5}$

# Chapter 13 :

# Geometry and Solid Figures

## Topics that you'll practice in this chapter:

- ✓ Angles
- ✓ Pythagorean Relationship
- ✓ Triangles
- ✓ Polygons
- ✓ Trapezoids
- ✓ Circles
- ✓ Cubes
- ✓ Rectangular Prism
- ✓ Cylinder
- ✓ Pyramids and Cone

*Geometry is the archetype of the beauty of the world.*

*Johannes Kepler*

# Angles

✍ **What is the value of $x$ in the following figures?**

1)

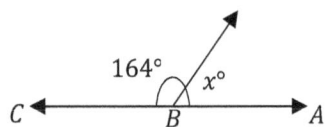

2)

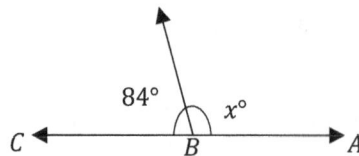

3)

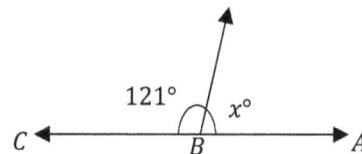

4)

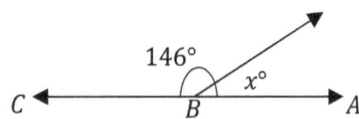

5)

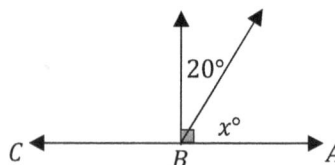

6)

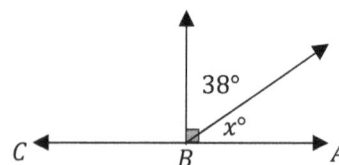

✍ **Calculate.**

7) Two supplement angles have equal measures. What is the measure of each angle? _____

8) The measure of an angle is seven fifth the measure of its supplement. What is the measure of the angle? _____

9) Two angles are complementary and the measure of one angle is 24 less than the other. What is the measure of the smaller angle?

_____

10) Two angles are complementary. The measure of one angle is one fifth the measure of the other. What is the measure of the bigger angle?

_____

11) Two supplementary angles are given. The measure of one angle is 40° less than the measure of the other. What does the smaller angle measure? _____

# Pythagorean Relationship

✎ **Do the following lengths form a right triangle?**

1)

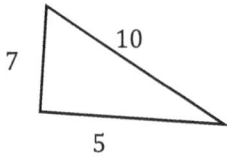

2)

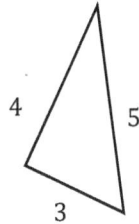

3)

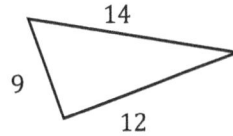

4)

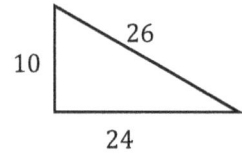

5)

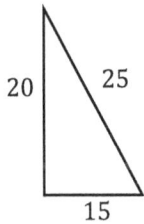

6)

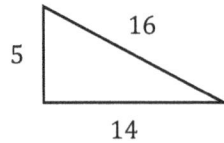

7)

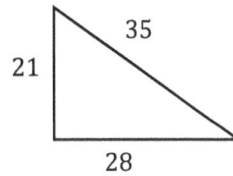

8)

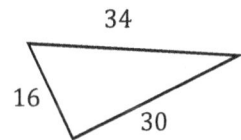

✎ **Find the missing side?**

9)

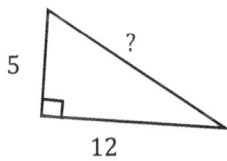

10)

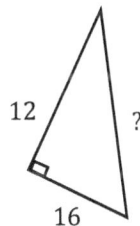

11)

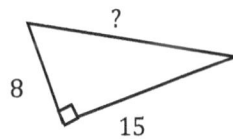

12)

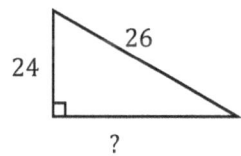

13)

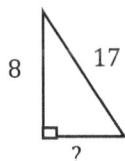

14)

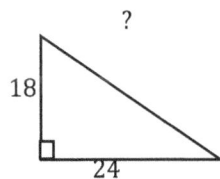

15)

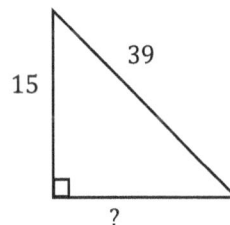

16)

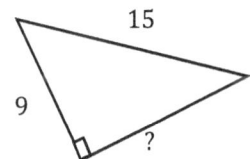

# Triangles

✎ **Find the measure of the unknown angle in each triangle.**

1)

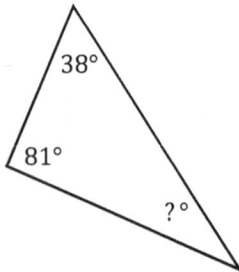

2)

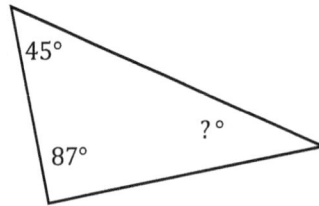

3)

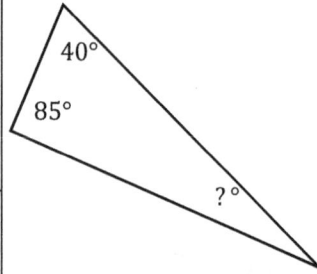

4)

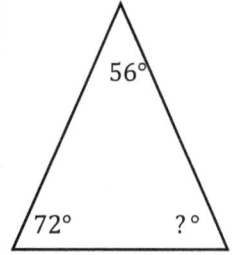

5)

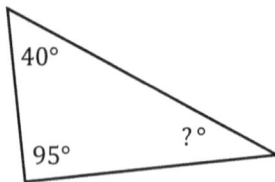

6)

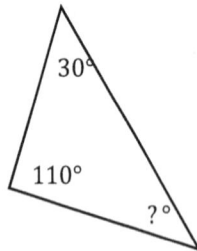

7)

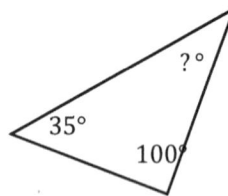

8)

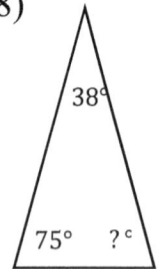

✎ **Find area of each triangle.**

9)

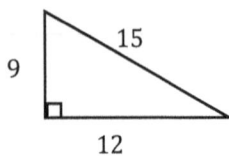

10)

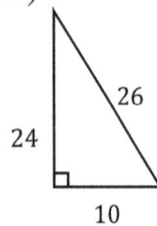

11)

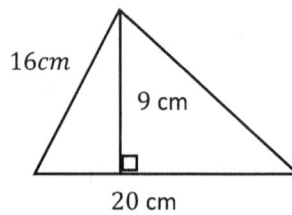

12)

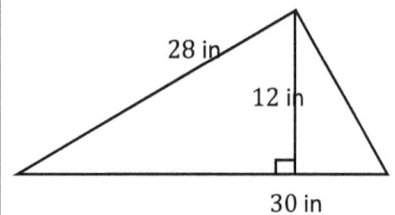

# Polygons

✎ **Find the perimeter of each shape.**

1)

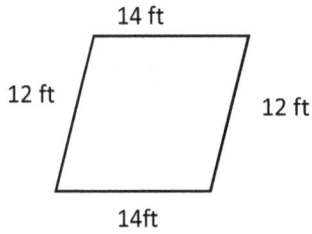

14 ft

12 ft        12 ft

14ft

2)

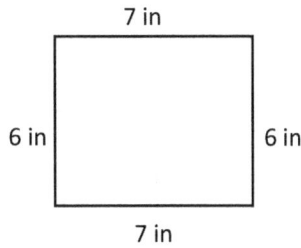

7 in

6 in        6 in

7 in

3)

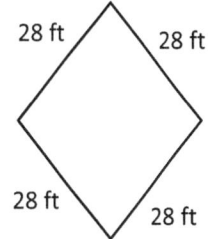

28 ft    28 ft

28 ft    28 ft

4) Square

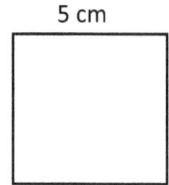

5 cm

5) Regular hexagon

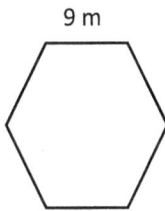

9 m

6)

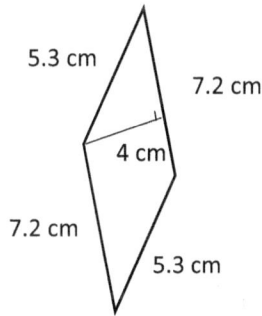

5.3 cm

7.2 cm

4 cm

7.2 cm

5.3 cm

7) Parallelogram

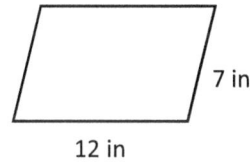

7 in

12 in

8) Square

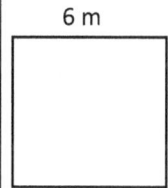

6 m

✎ **Find the area of each shape.**

9) Parallelogram

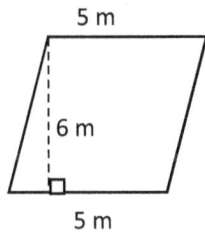

5 m

6 m

5 m

10) Rectangle

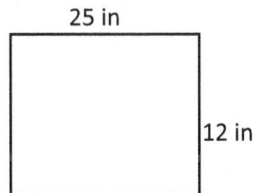

25 in

12 in

11) Rectangle

16 km

10 km

12) Square

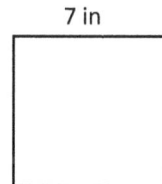

7 in

# Trapezoids

✍ **Find the area of each trapezoid.**

1)

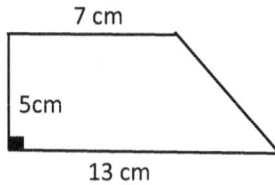

7 cm

5cm

13 cm

2)

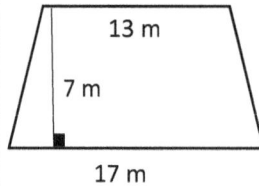

13 m

7 m

17 m

3)

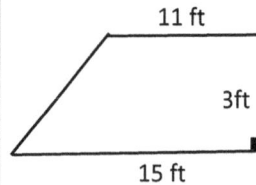

11 ft

3ft

15 ft

4)

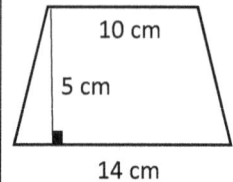

10 cm

5 cm

14 cm

5)

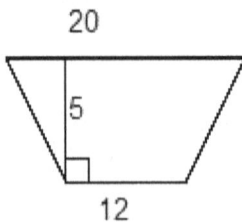

20

5

12

6)

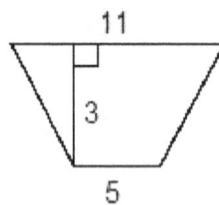

11

3

5

7)

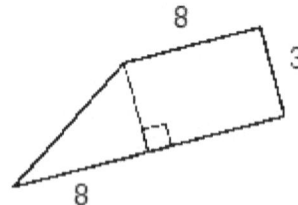

8

3

8

8)

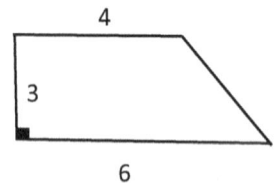

4

3

6

✍ **Calculate.**

1) A trapezoid has an area of 45 cm² and its height is 5 cm and one base is 5 cm. What is the other base length? _____

2) If a trapezoid has an area of 99 ft² and the lengths of the bases are 8 ft and 10 ft, find the height? _____

3) If a trapezoid has an area of 126 m² and its height is 14 m and one base is 6 m, find the other base length? _____

4) The area of a trapezoid is 440 ft² and its height is 22 ft. If one base of the trapezoid is 15 ft, what is the other base length?

_____

# Circles

✎ **Find the area of each circle.** ($\pi = 3.14$)

| 1) | 2) | 3) | 4) | 5) | 6) |
|---|---|---|---|---|---|
| 2.5 in | 5cm | 9 ft | 2 m | 11 cm | 10 miles |

| 7) | 8) | 9) | 10) | 11) | 12) |
|---|---|---|---|---|---|
| 13 in | 3 ft | 4 m | 12 cm | 7 miles | 8 ft |

✎ **Complete the table below.** ($\pi = 3.14$)

| Circle No. | Radius | Diameter | Circumference | Area |
|---|---|---|---|---|
| 1 | 1 $in$ | 2 $in$ | 6.28 $in$ | 3.14 $in^2$ |
| 2 | | 10 $m$ | | |
| 3 | | | | 28.26 $ft^2$ |
| 4 | | | 47.1 $mi$ | |
| 5 | | 11 $km$ | | |
| 6 | 7 $cm$ | | | |
| 7 | | 12 $ft$ | | |
| 8 | | | | 314 $m^2$ |
| 9 | | | 56.52 $in$ | |
| 10 | 4.5 $ft$ | | | |

# Cubes

✎ **Find the volume of each cube.**

| 1) | 2) | 3) | 4) | 5) | 6) |
|---|---|---|---|---|---|
| | 2 cm | 6 ft | 11 m | 13 in | 7 miles |

| 7) | 8 | 9) | 10) | 11) | 12) |
|---|---|---|---|---|---|
| 1.2 km | 9 cm | 2.1 ft | 12 mm | 0.2 in | 0.1 km |

✎ **Find the surface area of each cube.**

| 13) | 14) | 15) | 16) | 17) | 18) |
|---|---|---|---|---|---|
| | 7 m | 5 ft | 4.5 mm | 1.1 km | 11 cm |

# Rectangular Prism

✎ **Find the volume of each Rectangular Prism.**

1)

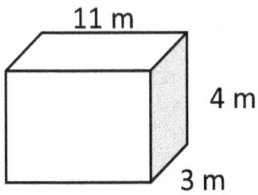

11 m
4 m
3 m

2)

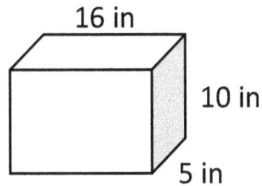

16 in
10 in
5 in

3)

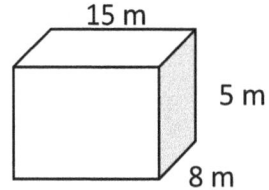

15 m
5 m
8 m

4)

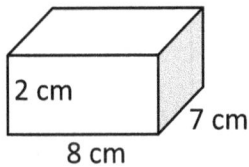

2 cm
7 cm
8 cm

5)

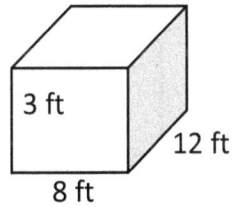

3 ft
12 ft
8 ft

6)

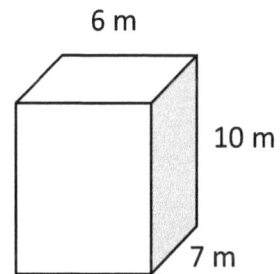

6 m
10 m
7 m

✎ **Find the surface area of each Rectangular Prism.**

7)

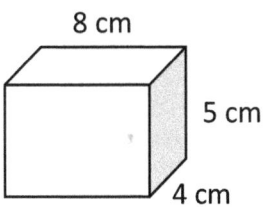

8 cm
5 cm
4 cm

8)

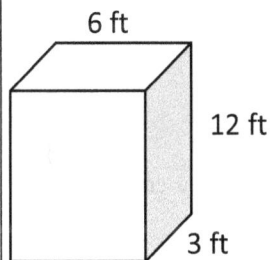

6 ft
12 ft
3 ft

9)

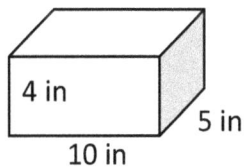

4 in
5 in
10 in

10)

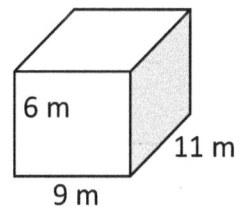

6 m
11 m
9 m

## Cylinder

✎ **Find the volume of each Cylinder. Round your answer to the nearest tenth.** ($\pi = 3.14$)

1)

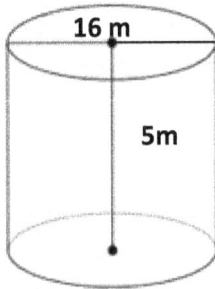

16 m

5m

2)

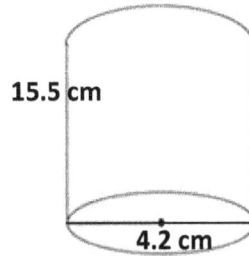

15.5 cm

4.2 cm

3)

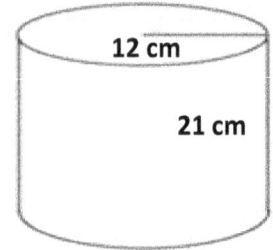

12 cm

21 cm

4)

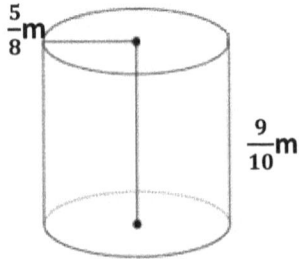

$\frac{5}{8}$m

$\frac{9}{10}$m

5)

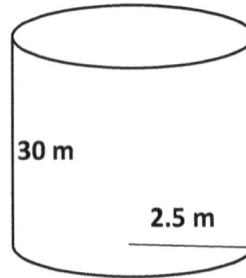

30 m

2.5 m

6)

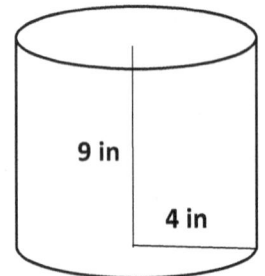

9 in

4 in

✎ **Find the surface area of each Cylinder.** ($\pi = 3.14$)

7)

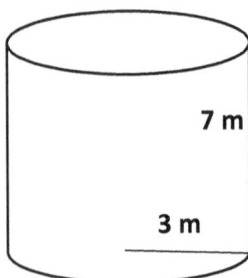

7 m

3 m

8)

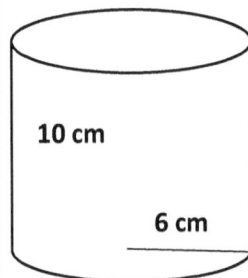

10 cm

6 cm

9)

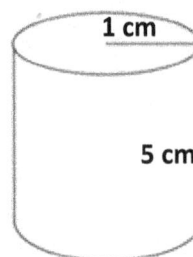

1 cm

5 cm

10)

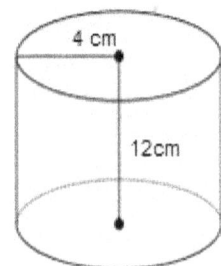

4 cm

12cm

# Pyramids and Cone

✎ **Find the volume of each Pyramid and Cone.** ($\pi = 3.14$)

1)

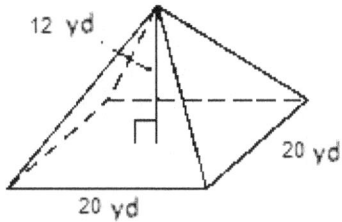

2)

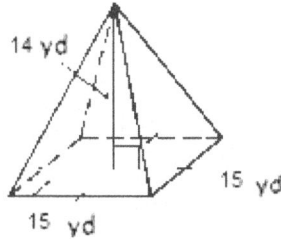

3)

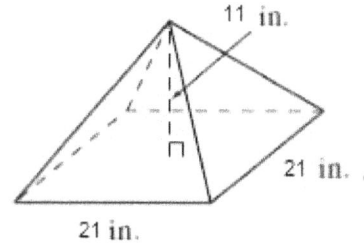

4)

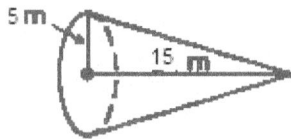

5)

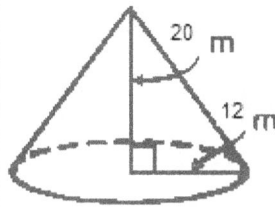

6)

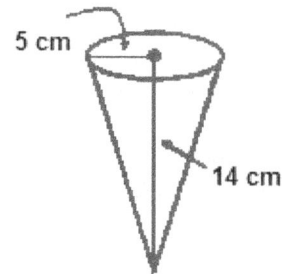

✎ **Find the surface area of each Pyramid and Cone.** ($\pi = 3.14$)

7)

8)

9)

10)

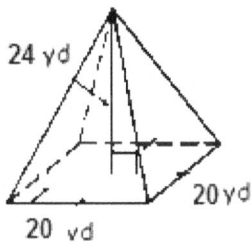

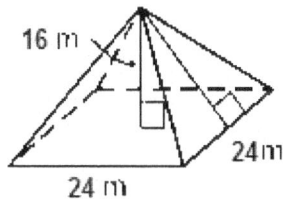

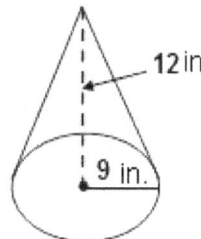

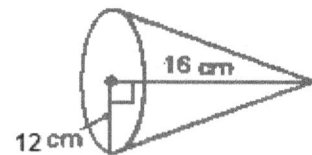

# Answers of Worksheets

**Angles**

| | | | |
|---|---|---|---|
| 1) 16° | 4) 34° | 7) 90° | 10) 75° |
| 2) 96° | 5) 70° | 8) 75° | 11) 70° |
| 3) 59° | 6) 52° | 9) 33° | |

**Pythagorean Relationship**

| | | | |
|---|---|---|---|
| 1) No | 5) Yes | 9) 13 | 13) 15 |
| 2) Yes | 6) No | 10) 20 | 14) 30 |
| 3) No | 7) Yes | 11) 17 | 15) 36 |
| 4) Yes | 8) Yes | 12) 10 | 16) 12 |

**Triangles**

| | | |
|---|---|---|
| 1) 60° | 5) 45° | 9) 54 $square\ unites$ |
| 2) 48° | 6) 40° | 10) 120 $square\ unites$ |
| 3) 55° | 7) 45° | 11) 90 $square\ unites$ |
| 4) 52° | 8) 67° | 12) 180 $square\ unites$ |

**Polygons**

| | | |
|---|---|---|
| 1) 52 $ft$ | 5) 54 $m$ | 9) 30 $m^2$ |
| 2) 26 $in$ | 6) 25 $cm$ | 10) 300 $in^2$ |
| 3) 112 $ft$ | 7) 38 $in$ | 11) 160 $km^2$ |
| 4) 20 $cm$ | 8) 24 $m$ | 12) 49 $in^2$ |

**Trapezoids**

| | | |
|---|---|---|
| 1) 50 $cm^2$ | 4) 60 $cm^2$ | 7) 36 |
| 2) 105 $m^2$ | 5) 80 | 8) 15 |
| 3) 39 $ft^2$ | 6) 24 | |

**Calculate**

| | | | |
|---|---|---|---|
| 1) 13 $cm$ | 2) 11 $ft$ | 3) 12 $m$ | 4) 25 $ft$ |

**Circles**

| | | |
|---|---|---|
| 1) 19.63 $in^2$ | 5) 379.94 $cm^2$ | 9) 12.56 $m^2$ |
| 2) 78.5 $cm^2$ | 6) 314 $miles^2$ | 10) 113.04 $cm^2$ |
| 3) 254.34 $ft^2$ | 7) 132.67 $in^2$ | 11) 38.47 $miles^2$ |
| 4) 12.56 $m^2$ | 8) 7.07 $ft^2$ | 12) 50.24 $ft^2$ |

| Circle No. | Radius | Diameter | Circumference | Area |
|---|---|---|---|---|
| 1 | 1 in | 2 in | 6.28 in | $3.14\ in^2$ |
| 2 | 5 m | 10 m | 31.4 m | $78.5\ m^2$ |
| 3 | 3 ft | 6 ft | 18.84 ft | $28.26\ ft^2$ |
| 4 | 7.5 miles | 15 mi | 47.1 mi | $176.63\ mi^2$ |
| 5 | 5.5 km | 11 km | 34.54 km | $94.99\ km^2$ |
| 6 | 7 cm | 14 cm | 43.96 cm | $153.86\ cm^2$ |
| 7 | 6 ft | 12 ft | 37.68 feet | $113.04\ ft^2$ |
| 8 | 10 m | 20 m | 62.8 m | $314\ m^2$ |
| 9 | 9 in | 18 in | 56.52 in | $254.34\ in^2$ |
| 10 | 4.5 ft | 9 ft | 28.26 ft | $63.585\ ft^2$ |

## Cubes

1) 12
2) $8\ cm^3$
3) $216\ ft^3$
4) $1,331\ m^3$
5) $2,197\ in^3$

6) $343\ miles^3$
7) $1.728\ km^3$
8) $729\ cm^3$
9) $9.261\ ft^3$
10) $1,728\ mm^3$

11) $0.008\ in^3$
12) $0.001\ km^3$
13) 27
14) $294\ m^2$
15) $150\ ft^2$

16) $121.5\ mm^2$
17) $7.26\ km^2$
18) $726\ cm^2$

## Rectangular Prism

1) $132\ m^3$
2) $800\ in^3$
3) $600\ m^3$

4) $112\ cm^3$
5) $288\ ft^3$
6) $420\ m^3$

7) $184\ cm^2$
8) $252\ ft^2$
9) $220\ in^2$

10) $438\ m^2$

## Cylinder

1) $1,004.8\ m^3$
2) $214.6\ cm^3$
3) $9,495.4\ cm^3$

4) $1.1\ m^3$
5) $588.8\ m^3$
6) $452.2\ in^3$

7) $188.4\ m^2$
8) $602.9\ cm^2$
9) $37.7\ cm^2$

10) $401.9\ m^2$

## Pyramids and Cone

1) $1,600\ yd^3$
2) $1,050\ yd^3$
3) $1,617\ in^3$

4) $392.5\ m^3$
5) $3,014.4\ m^3$
6) $366.33\ cm^3$

7) $1,440\ yd^2$
8) $1,536\ m^2$
9) $678.24\ in^2$

10) $1,205.76\ cm^2$

# Chapter 14 :

# Trigonometric Functions

## Topics that you'll practice in this chapter:

✓ Trig ratios of General Angles

✓ Sketch Each Angle in Standard Position

✓ Finding Co–Terminal Angles and Reference Angles

✓ Angles in Radians

✓ Angles in Degrees

✓ Evaluating Each Trigonometric Expression

✓ Missing Sides and Angles of a Right Triangle

✓ Arc Length and Sector Area

*Mathematics is like checkers in being suitable for the young, not too difficult, amusing, and without peril to the state.* — *Plato*

# Trig ratios of General Angles

✎ **Evaluate.**

1) $\sin 135° = $ _____

2) $\sin 300° = $ _____

3) $\cos - 225° = $ _____

4) $\cos 270° = $ _____

5) $\sin 450° = $ _____

6) $\sin -330° = $ _____

7) $\tan 60° = $ _____

8) $\cot 180° = $ _____

9) $\tan 240° = $ _____

10) $\cot 90° = $ _____

11) $\sec 180° = $ _____

12) $\csc 90° = $ _____

13) $\cot -270° = $ _____

14) $\sec 360° = $ _____

15) $\cos - 45° = $ _____

16) $\sec 120° = $ _____

17) $\csc 360° = $ _____

18) $\cot -45° = $ _____

✎ **Find the exact value of each trigonometric function. Some may be undefined.**

19) $\sec 2\pi = $ _____

20) $\tan -\dfrac{5\pi}{2} = $ _____

21) $\cos \dfrac{11\pi}{2} = $ _____

22) $\cot \dfrac{9\pi}{4} = $ _____

23) $\sec - 6\pi = $ _____

24) $\sec \dfrac{\pi}{4} = $ _____

25) $\csc \dfrac{8\pi}{3} = $ _____

26) $\cot \dfrac{10\pi}{3} = $ _____

27) $\csc -\dfrac{\pi}{2} = $ _____

28) $\cot \dfrac{2\pi}{3} = $ _____

# Sketch Each Angle in Standard Position

✎ **Draw each angle with the given measure in standard position.**

1) −570°

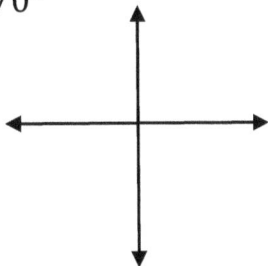

4) −690°

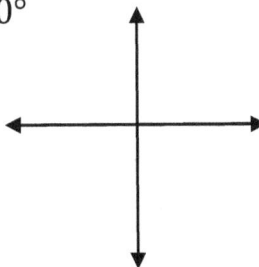

2) 750°

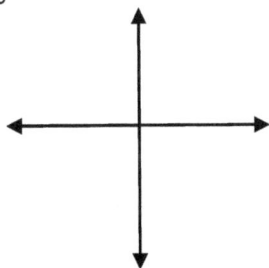

5) $\frac{13\pi}{6}$

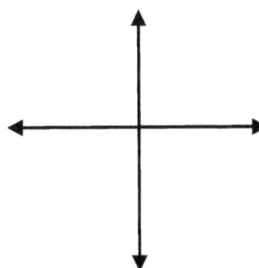

3) 1,110°

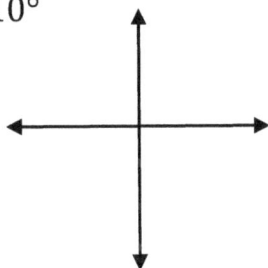

6) $-\frac{11\pi}{6}$

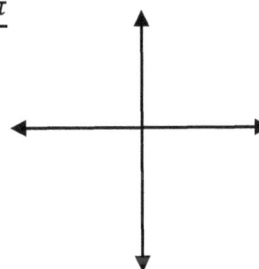

# Finding Co-terminal Angles and Reference Angles

✎ Find a conterminal angle between 0° and 360° for each angle provided.

1) $-315° =$

3) $-225° =$

2) $-210° =$

4) $-540° =$

✎ Find a conterminal angle between 0 and 2π for each given angle.

5) $\dfrac{18\pi}{5} =$

7) $-\dfrac{13\pi}{4} =$

6) $-\dfrac{19\pi}{6} =$

8) $\dfrac{14\pi}{3} =$

✎ Find the reference angle of each angle.

9)

10)

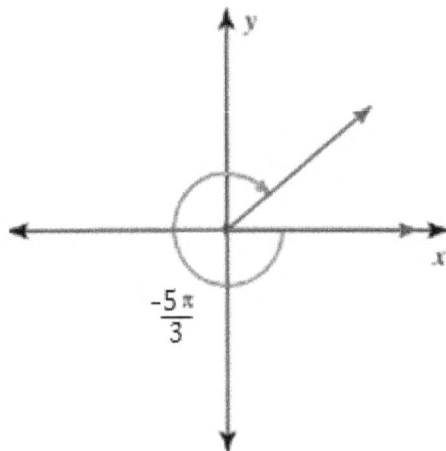

$\dfrac{-5\pi}{3}$

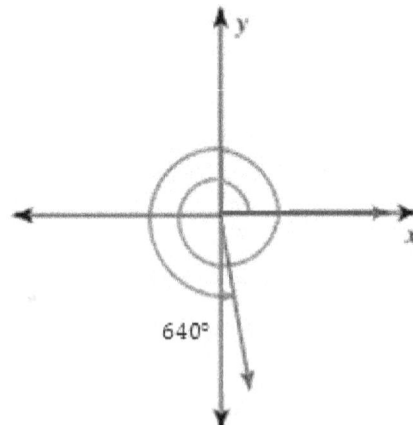

640°

# Angles and Angle Measure

✎ **Convert each degree measure into radians.**

1) $216° = $ ___

2) $660° = $ ___

3) $420° = $ ___

4) $220° = $ ___

5) $210° = $ ___

6) $270° = $ ___

7) $-300° = $ ___

8) $810° = $ ___

9) $330° = $ ___

10) $140° = $ ___

11) $480° = $ ___

12) $405° = $ ___

13) $-450° = $ ___

14) $-126° = $ ___

15) $-675° = $ ___

16) $150° = $ ___

17) $-468° = $ ___

18) $340° = $ ___

19) $-440° = $ ___

20) $342° = $ ___

21) $230° = $ ___

✎ **Convert each radian measure into degrees.**

22) $\frac{\pi}{10} = $

23) $\frac{5\pi}{12} = $

24) $\frac{7\pi}{3} = $

25) $\frac{3\pi}{20} = $

26) $-\frac{6\pi}{5} = $

27) $\frac{11\pi}{18} = $

28) $-\frac{14\pi}{5} = $

29) $\frac{5\pi}{18} = $

30) $\frac{7\pi}{36} = $

31) $\frac{17\pi}{18} = $

32) $-\frac{13\pi}{30} = $

33) $\frac{7\pi}{9} = $

34) $-\frac{19\pi}{18} = $

35) $\frac{7\pi}{60} = $

36) $-\frac{3\pi}{10} = $

37) $\frac{11\pi}{30} = $

38) $-\frac{2\pi}{9} = $

39) $-\frac{7\pi}{10} = $

# Evaluating Trigonometric Functions

✍ **Find the exact value of each trigonometric function.**

1) $\cos 780° =$ _____

2) $\tan \dfrac{5\pi}{3} =$ _____

3) $\tan -\dfrac{\pi}{6} =$ _____

4) $\cot -\dfrac{9\pi}{4} =$ _____

5) $\cos -\dfrac{7\pi}{6} =$ _____

6) $\cos 135° =$ _____

7) $\sin 240° =$ _____

8) $\tan 330° =$ _____

9) $\cot 420° =$ _____

10) $\tan -495° =$ _____

11) $\cot 315° =$ _____

12) $\sin -240° =$ _____

13) $\cot 225° =$ _____

✍ **Use the given point on the terminal side of angle θ to find the value of the trigonometric function indicated.**

14) $\sin\theta, \ (-6, 8)$

15) $\cos\theta, \ (-6, 8)$

16) $\sec\theta, \ (3, \ 5)$

17) $\cos\theta, \ (10, 24)$

18) $\sin\theta, \ (6, -6)$

19) $\tan\theta, \ (-2, -\sqrt{12})$

## Missing Sides and Angles of a Right Triangle

✍ Find the value of each trigonometric ratio as fractions in their simplest form.

1) $cot\ x$

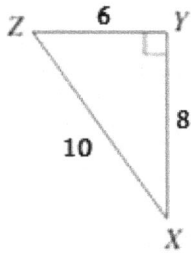

2) $cos\ A$

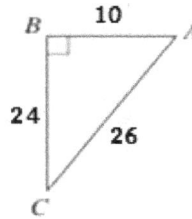

✍ Find the missing sides. Round answers to the nearest tenth.

3)

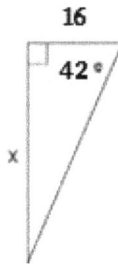

4)

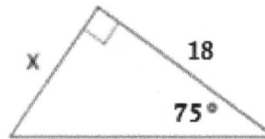

5)

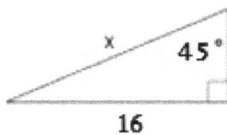

6)

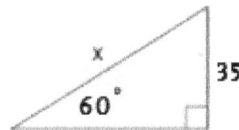

# Arc Length and Sector Area

✍ **Find the length of each arc. Round your answers to the nearest tenth.**

($\pi = 3.14$)

1) $r = 28$ cm, $\theta = 30°$

2) $r = 14$ ft, $\theta = 95°$

3) $r = 22$ ft, $\theta = 50°$

4) $r = 16\,m$, $\theta = 85°$

✍ **Find area of each sector. Do *not* round. Round your answers to the nearest tenth.** ($\pi = 3.14$)

5)

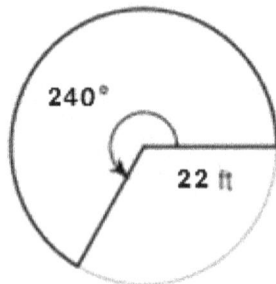

6)

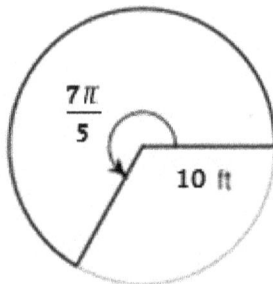

7)

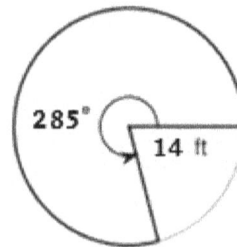

8)

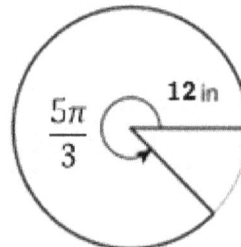

# Answers of Worksheets

## Trig Ratios of General Angles

1) $\frac{\sqrt{2}}{2}$

2) $-\frac{\sqrt{3}}{2}$

3) $-\frac{\sqrt{2}}{2}$

4) 0

5) 1

6) $\frac{1}{2}$

7) $\sqrt{3}$

8) Undefined

9) $\sqrt{3}$

10) 0

11) −1

12) 1

13) 0

14) 1

15) $\frac{\sqrt{2}}{2}$

16) −2

17) Undefined

18) −1

19) 1

20) Undefined

21) 0

22) 1

23) 1

24) $\sqrt{2}$

25) $\frac{2\sqrt{3}}{3}$

26) $\frac{\sqrt{3}}{3}$

27) −1

28) $-\frac{\sqrt{3}}{3}$

## Sketch Each Angle in Standard Position

1) −570°

2) 750°

3) 1,110°

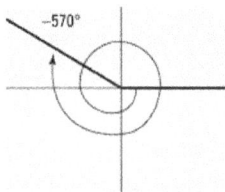
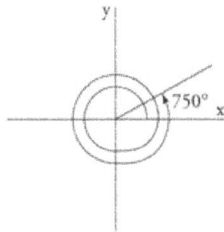
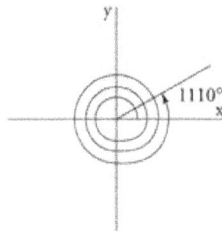

4) −690°

5) $\frac{13\pi}{6} = 390°$

6) $-\frac{11\pi}{6} = -330°$

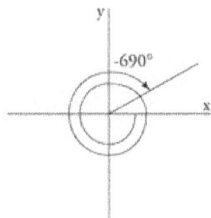
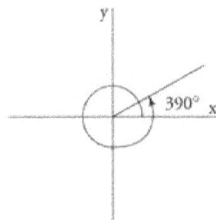
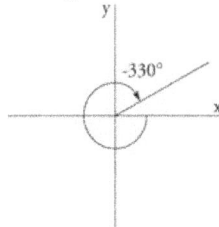

## Finding Co–Terminal Angles and Reference Angles

1) 45°

2) 150°

3) 135°

4) 180°

5) $\frac{4\pi}{5}$

6) $\frac{5\pi}{6}$

7) $\frac{3\pi}{4}$

8) $\frac{2\pi}{3}$

9) $\frac{\pi}{3}$

10) 80°

## Angles and Angle Measure

1) $\frac{6\pi}{5}$

2) $\frac{11\pi}{3}$

3) $\frac{7\pi}{3}$

4) $\frac{11\pi}{9}$

5) $\frac{7\pi}{6}$

6) $\frac{3\pi}{2}$

7) $-\frac{5\pi}{3}$

8) $\frac{9\pi}{2}$

9) $\frac{11\pi}{6}$

10) $\frac{7\pi}{9}$

11) $\frac{8\pi}{3}$

12) $\frac{9\pi}{4}$

13) $-\frac{5}{2}\pi$

14) $-\frac{7\pi}{10}$

15) $-\frac{15\pi}{4}$

16) $\frac{5\pi}{6}$

17) $-\frac{13\pi}{5}$

18) $\frac{17\pi}{9}$

19) $-\frac{22\pi}{9}$

20) $\frac{19\pi}{10}$

21) $\frac{23\pi}{18}$

22) $18°$

23) $75°$

24) $420°$

25) $27°$

26) $-216°$

27) $110°$

28) $-504°$

29) $50°$

30) $35°$

31) $170°$

32) $-78°$

33) $140°$

34) $-190°$

35) $21°$

36) $-54°$

37) $66°$

38) $-40°$

39) $-126°$

**Evaluating Each Trigonometric Functions**

1) $\frac{1}{2}$

2) $-\sqrt{3}$

3) $-\frac{\sqrt{3}}{3}$

4) $-1$

5) $-\frac{\sqrt{3}}{2}$

6) $-\frac{\sqrt{2}}{2}$

7) $-\frac{\sqrt{3}}{2}$

8) $-\frac{\sqrt{3}}{3}$

9) $\frac{\sqrt{3}}{3}$

10) $1$

11) $-1$

12) $\frac{\sqrt{3}}{2}$

13) $1$

14) $0.8$

15) $-0.6$

16) $\frac{\sqrt{34}}{5}$

17) $\frac{5}{13}$

18) $-\frac{\sqrt{2}}{2}$

19) $\sqrt{3}$

**Missing Sides and Angles of a Right Triangle**

1) $\frac{4}{3}$

2) $\frac{5}{13}$

3) $14.4$

4) $67.2$

5) $22.6$

6) $40.4$

**Arc Length and Sector Area**

1) $14.7\ cm$

2) $23.2\ ft$

3) $19.2\ ft$

4) $23.7m$

5) $1,013.7\ ft^2$

6) $220\ in^2$

7) $487.5\ ft^2$

8) $377\ in^2$

# Chapter 15 :

# Statistics and Probability

## Topics that you'll practice in this chapter:

- ✓ Mean and Median
- ✓ Mode and Range
- ✓ Histograms
- ✓ Stem–and–Leaf Plot
- ✓ Pie Graph
- ✓ Probability Problems
- ✓ Factorials
- ✓ Combinations and Permutation

*"The book of nature is written in the language of Mathematic."*

*- Galileo.*

# Mean and Median

✎ **Find Mean and Median of the Given Data.**

1) 8, 7, 14, 4, 8

2) 14, 8, 25, 19, 16, 33, 11

3) 23, 18, 15, 12, 17

4) 34, 14, 10, 15, 6, 11

5) 10, 19, 6, 8, 32, 20, 17

6) 17, 26, 39, 69, 20, 6

7) 40, 38, 18, 11, 9, 2, 7, 32, 41

8) 24, 21, 31, 12, 33, 32, 22

9) 16, 14, 20, 41, 15, 20, 38, 4

10) 20, 20, 30, 18, 6, 28, 12, 46

11) 12, 7, 10, 11, 16, 22

12) 10, 29, 27, 12, 2, 15, 10, 3

✎ **Calculate.**

13) In a javelin throw competition, five athletics score 56, 34, 62, 23 and 19 meters. What are their Mean and Median? _____

14) Eva went to shop and bought 8 apples, 14 peaches, 6 bananas, 4 pineapples and 12 melons. What are the Mean and Median of her purchase? _____

15) Bob has 17 black pen, 19 red pen, 14 green pens, 20 blue pens and 5 boxes of yellow pens. If the Mean and Median are 19 respectively, what is the number of yellow pens in each box? _____

# Mode and Range

✍ **Find Mode and Rage of the Given Data.**

1) 4, 3, 7, 3, 3, 4

    Mode: _____       Range: _____

2) 18, 18, 24, 26, 18, 8, 14, 22

    Mode: _____       Range: _____

3) 8, 8, 8, 16, 19, 22, 20, 9, 13

    Mode: _____       Range: _____

4) 24, 24, 14, 28, 20, 18, 20, 24

    Mode: _____       Range: _____

5) 6, 21, 27, 24, 27, 27

    Mode: _____       Range: _____

6) 21, 8, 8, 7, 8, 12, 10, 22, 18, 13

    Mode: _____       Range: _____

7) 7, 4, 4, 6, 13, 13, 13, 0, 2, 2

    Mode: _____       Range: _____

8) 5, 8, 5, 14, 12, 14, 3, 5, 18

    Mode: _____       Range: _____

9) 7, 7, 7, 12, 7, 3, 8, 16, 3, 17

    Mode: _____       Range: _____

10) 15, 15, 19, 16, 4, 16, 10, 15

    Mode: _____       Range: _____

11) 6, 6, 5, 6, 42, 13, 19, 2

    Mode: _____       Range: _____

12) 8, 8, 9, 8, 9, 4, 34, 22

    Mode: _____       Range: _____

✍ **Calculate.**

13) A stationery sold 12 pencils, 56 red pens, 24 blue pens, 20 notebooks, 12 erasers, 21 rulers and 11 color pencils. What are the Mode and Range for the stationery sells?

                Mode: _____       Range: _____

14) In an English test, eight students score 10, 15, 15, 18 18, 16, 15 and 15. What are their Mode and Range? _____

15) What is the range of the first 6 even numbers greater than 8?

    _____

# Times Series

✎ **Use the following Graph to complete the table.**

| Day | Distance (km) |
|-----|---------------|
| 1 | |
| 2 | |
| | |
| | |
| | |
| | |

Distance

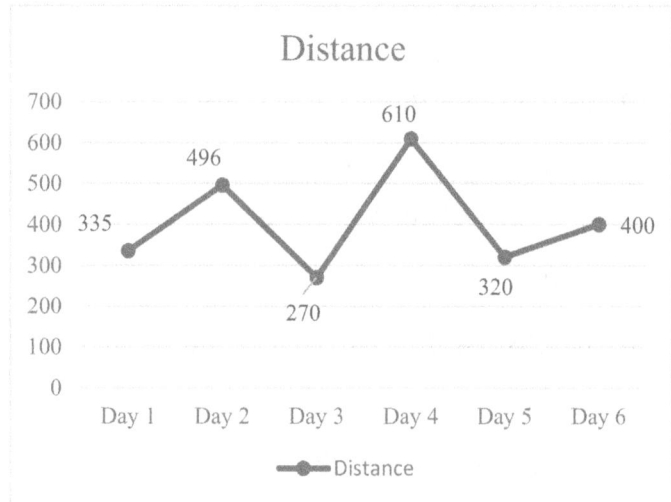

The following table shows the number of births in the US from 2007 to 2012 (in millions).

| Year | Number of births (in millions) |
|------|-------------------------------|
| 2007 | 4.15 |
| 2008 | 3.70 |
| 2009 | 3.45 |
| 2010 | 3.20 |
| 2011 | 1.75 |
| 2012 | 2.98 |

Draw a Time Series for the table.

## Stem–and–Leaf Plot

🖎**Make stem ad leaf plots for the given data.**

1) 24, 26, 29, 20, 53, 27, 51, 55, 36, 21, 37, 30

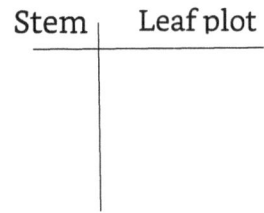

Stem | Leaf plot

2) 11, 59, 66, 14, 18, 19, 59, 65, 69, 61, 68, 65

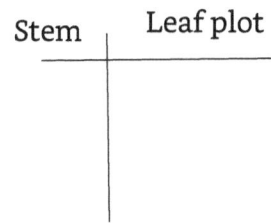

Stem | Leaf plot

3) 121, 55, 66, 54, 112, 128, 63, 125, 59, 123, 68, 119

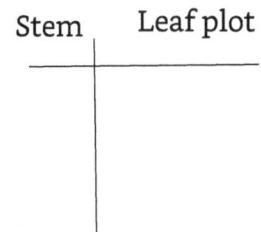

Stem | Leaf plot

4) 51, 32, 100, 56, 84, 36, 107, 56, 85, 39, 56, 106, 89

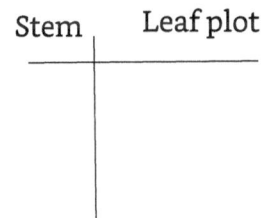

Stem | Leaf plot

5) 33, 89, 19, 87, 81, 16, 11, 30, 86, 35, 17, 35, 13

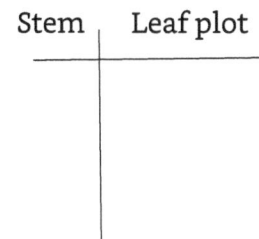

Stem | Leaf plot

6) 60, 92, 22, 25, 67, 93, 95, 62, 21, 64, 98, 29

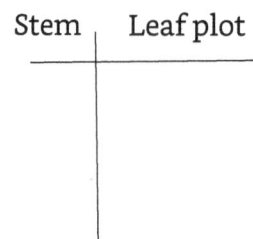

Stem | Leaf plot

# Pie Graph

The circle graph below shows all Robert's expenses for last month. Robert spent $140 on his hobbies last month.

Answer following questions based on the Pie graph.

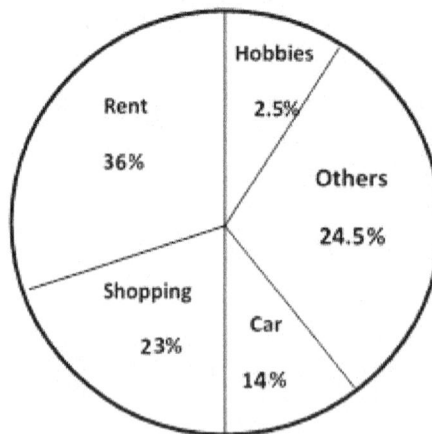

1) How much was Robert's total expenses last month? _____

2) How much did Robert spend on his car last month? _____

3) How much did Robert spend for shopping last month? _____

4) How much did Robert spend on his rent last month? _____

5) What fraction is Robert's expenses for his rent and car out of his total

   expenses last month? _____

## Probability Problems

🖎**Calculate.**

1) A number is chosen at random from 1 to 10. Find the probability of selecting number 6 or smaller numbers. _____

2) Bag A contains 18 red marbles and 6 green marbles. Bag B contains 16 black marbles and 8 orange marbles. What is the probability of selecting a green marble at random from bag A? What is the probability of selecting a black marble at random from Bag B? _____

3) A number is chosen at random from 1 to 20. What is the probability of selecting multiples of 4? _____

4) A card is chosen from a well-shuffled deck of 52 cards. What is the probability that the card will be a queen? _____

5) A number is chosen at random from 1 to 15. What is the probability of selecting a multiple of 3 or 5? _____

A spinner numbered 1–8, is spun once. What is the probability of spinning …?

6) an Odd number? _____     7) a multiple of 2? _____

8) a multiple of 5? _____     9) number 10? _____

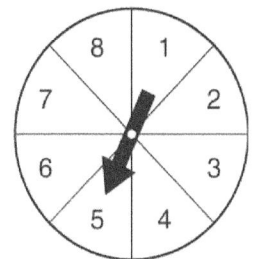

# Factorials

✎ **Determine the value for each expression.**

1) $4! + 0! =$

2) $2! + 5! =$

3) $(2!)^2 =$

4) $5! - 3! =$

5) $6! - 3! + 10 =$

6) $3! \times 4 - 15 =$

7) $(2! + 3!)^2 =$

8) $(4! - 3!)^2 =$

9) $(3! \, 0!)^2 - 10 =$

10) $\dfrac{10!}{8!} =$

11) $\dfrac{6!}{4!} =$

12) $\dfrac{6!}{5!} =$

13) $\dfrac{15!}{13!} =$

14) $\dfrac{n!}{(n-3)!} =$

15) $\dfrac{(n+2)!}{n!} =$

16) $\dfrac{(2+2!)^3}{2!} =$

17) $\dfrac{5(n+2)!}{(n+1)!} =$

18) $\dfrac{22!}{20!4!} =$

19) $\dfrac{13!}{11!3!} =$

20) $\dfrac{9 \times 210!}{3(7 \times 30)!} =$

21) $\dfrac{32!}{31!2!} =$

22) $\dfrac{11!12!}{10!13!} =$

23) $\dfrac{16!15!}{14!14!} =$

24) $\dfrac{(5 \times 3)!}{0!14!} =$

25) $\dfrac{4!(5n-2)!}{(5n)!} =$

26) $\dfrac{4n(4n+7)!}{(4n+8)!} =$

27) $\dfrac{(n-2)!(n+1)}{(n+2)!} =$

# Combinations and Permutations

✑ **Calculate the value of each.**

1) $6! =$ _____

2) $2! \times 5! =$ _____

3) $3 \times 4! =$ _____

4) $5! + 3! =$ _____

5) $7! =$ _____

6) $4! =$ _____

7) $3! + 3! =$ _____

8) $7! - 5! =$ _____

✑ **Find the answer for each word problems.**

9) Susan is baking cookies. She uses sugar, butter, Vanilla, eggs and flour. How many different orders of ingredients can she try? _____

10) Albert is planning for his vacation. He wants to go to museum, watch a movie, go to the beach, play the game and play football. How many ways of ordering are there for him? _____

11) How many 4-digit numbers can be named using the digits 3, 4, 5, and 6 without repetition? _____

12) In how many ways can 5 boys be arranged in a straight line? _____

13) In how many ways can 6 athletes be arranged in a straight line? _____

14) A professor is going to arrange her 7 students in a straight line. In how many ways can she do this? _____

15) How many code symbols can be formed with the letters for the word GAMES? _____

16) In how many ways a team of 7 basketball players can choose a captain and co-captain? _____

# Answers of Worksheets

### Mean and Median

1) Mean: 8.2, Median: 8
2) Mean: 18, Median: 16
3) Mean: 17, Median: 17
4) Mean: 15, Median: 12.5
5) Mean: 16, Median: 17

6) Mean: 29.5, Median: 23
7) Mean: 22, Median: 18
8) Mean: 25, Median: 24
9) Mean: 21, Median: 18
10) Mean: 22.5, Median: 20

11) Mean: 13, Median: 11.5
12) Mean: 13.5, Median: 11
13) Mean: 38.8, Median: 34
14) Mean: 8.8, Median: 8
15) 5

### Mode and Range

1) Mode: 3, Range: 4
2) Mode: 18, Range: 18
3) Mode: 8, Range: 14
4) Mode: 24, Range: 14
5) Mode: 27, Range: 21

6) Mode: 8, Range: 15
7) Mode: 13, Range: 13
8) Mode: 5, Range: 15
9) Mode: 7, Range: 14
10) Mode: 15, Range: 15

11) Mode: 6, Range: 40
12) Mode: 8, Range: 30
13) Mode: 12, Range: 45
14) Mode: 15, Range: 8
15) 10

### Time series

| Day | Distance (km) |
|-----|---------------|
| 1 | 335 |
| 2 | 496 |
| 3 | 270 |
| 4 | 610 |
| 5 | 320 |
| 6 | 400 |

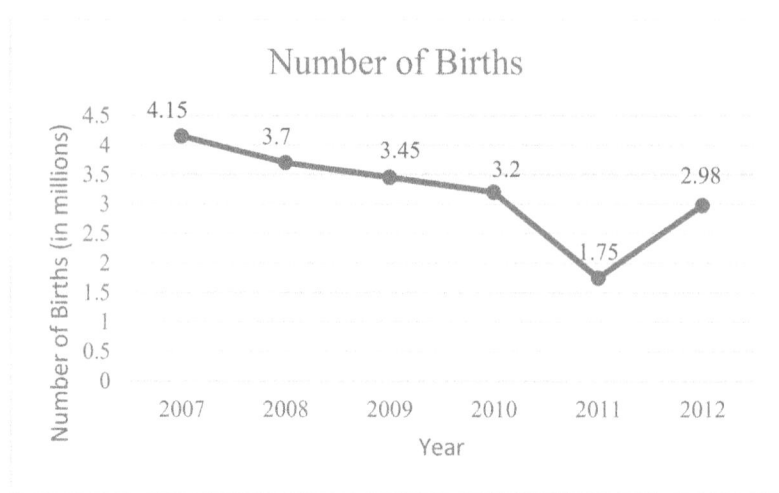

Number of Births

### Stem–And–Leaf Plot

1)

| Stem | leaf |
|------|------|
| 2 | 0 1 4 6 7 9 |
| 3 | 0 6 7 |
| 5 | 1 3 5 |

2)

| Stem | leaf |
|------|------|
| 1 | 1 4 8 9 |
| 5 | 9 9 |
| 6 | 1 5 5 6 8 9 |

3)

| Stem | leaf |
|------|------|
| 5 | 4 5 9 |
| 6 | 3 6 8 |
| 11 | 2 9 |
| 12 | 1 3 5 8 |

4)

| Stem | leaf |
|------|------|
| 3 | 2 6 9 |
| 5 | 1 6 6 6 |
| 8 | 4 5 9 |
| 10 | 0 6 7 |

5)

| Stem | leaf |
|------|------|
| 1 | 1 3 6 7 9 |
| 3 | 0  3 5 5 |
| 8 | 1 6 7 9 |

6)

| Stem | leaf |
|------|------|
| 2 | 2 1 5 9 |
| 6 | 0 2 4 7 |
| 9 | 2 3 5 8 |

## Pie Graph

1) $5,600

2) $784

3) $1,288

4) $2,016

5) $\frac{1}{2}$

## Probability Problems

1) $\frac{3}{5}$

2) $\frac{1}{4}, \frac{2}{3}$

3) $\frac{1}{4}$

4) $\frac{1}{13}$

5) $\frac{7}{15}$

6) $\frac{1}{2}$

7) $\frac{1}{2}$

8) $\frac{1}{8}$

9) 0

## Factorials

1) 25

2) 122

3) 4

4) 114

5) 724

6) 9

7) 64

8) 324

9) 26

10) 90

11) 30

12) 6

13) 210

14) $n(n-1)(n-2)$

15) $(n+1)(n+2)$

16) 32

17) $5(n+2)$

18) 19.25

19) 26

20) 3

21) 16

22) $\frac{11}{13}$

23) 3,600

24) 15

25) $\frac{24}{5n(5n-1)}$

26) $\frac{n}{(n+2)}$

27) $\frac{1}{n(n-1)(n+2)}$

## Combinations and Permutations

1) 720

2) 240

3) 72

4) 126

5) 5,040

6) 24

7) 12

8) 4,920

9) 120

10) 120

11) 24

12) 120

13) 720

14) 5,040

15) 120

16) 42

# Chapter 16 :
# PSAT Math Test Review

The Preliminary PSAT/ National Merit Scholarship Qualifying Test (PSAT/NMSQT) is a standardized test used for college admissions in the United States. 10th and 11th graders take the PSAT to practice for the PSAT and to secure a National Merit distinction or scholarship.

The PSAT is similar to the PSAT in both format and content. There are three sections on the PSAT:

- Reading
- Math
- Writing

The PSAT Math section is divided into two subsections:

A **No Calculator Section** contains 17 questions and students cannot use a calculator. Students have 25 minutes to complete this section.

A **Calculator Section** contains 31 questions. Students have 45 minutes to complete this section.

40 questions are multiple choice questions, and 8 questions are grid-ins.

PSAT Math cover the following topics:

- Pre-Algebra
- Algebra
- Coordinate Geometry
- Plane Geometry
- Date analysis and basic Statistics
- Trigonometry

In this section, there are two complete PSAT Math Tests. Take these tests to see what score you'll be able to receive on a real PSAT test.

*The hardest arithmetic to master is that which enables us to count our blessings.*

   *⁻Eric Hoffer*

# Time to Test

**Time to refine your skill with a practice examination.**

Take a practice PSAT Math Test to simulate the test day experience. After you've finished, score your test using the answer key.

## Before You Start

- You'll need a pencil, a calculator and a timer to take the test.
- For each question, there are four possible answers. Choose which one is best.
- After you've finished the test, review the answer key to see where you went wrong.

## Good Luck!

# PSAT Math Practice Test Answer Sheets

Remove (or photocopy) these answer sheets and use them to complete the practice tests.

## PSAT Practice Test - Section 1

1  Ⓐ Ⓑ Ⓒ Ⓓ     6  Ⓐ Ⓑ Ⓒ Ⓓ     11  Ⓐ Ⓑ Ⓒ Ⓓ

2  Ⓐ Ⓑ Ⓒ Ⓓ     7  Ⓐ Ⓑ Ⓒ Ⓓ     12  Ⓐ Ⓑ Ⓒ Ⓓ

3  Ⓐ Ⓑ Ⓒ Ⓓ     8  Ⓐ Ⓑ Ⓒ Ⓓ     13  Ⓐ Ⓑ Ⓒ Ⓓ

4  Ⓐ Ⓑ Ⓒ Ⓓ     9  Ⓐ Ⓑ Ⓒ Ⓓ

5  Ⓐ Ⓑ Ⓒ Ⓓ     10  Ⓐ Ⓑ Ⓒ Ⓓ

14   15   16

17

## PSAT Practice Test - Section 2

| | | | | | | |
|---|---|---|---|---|---|---|
| 1 | Ⓐ Ⓑ Ⓒ Ⓓ | 8 | Ⓐ Ⓑ Ⓒ Ⓓ | 15 | Ⓐ Ⓑ Ⓒ Ⓓ | 22 Ⓐ Ⓑ Ⓒ Ⓓ |
| 2 | Ⓐ Ⓑ Ⓒ Ⓓ | 9 | Ⓐ Ⓑ Ⓒ Ⓓ | 16 | Ⓐ Ⓑ Ⓒ Ⓓ | 23 Ⓐ Ⓑ Ⓒ Ⓓ |
| 3 | Ⓐ Ⓑ Ⓒ Ⓓ | 10 | Ⓐ Ⓑ Ⓒ Ⓓ | 17 | Ⓐ Ⓑ Ⓒ Ⓓ | 24 Ⓐ Ⓑ Ⓒ Ⓓ |
| 4 | Ⓐ Ⓑ Ⓒ Ⓓ | 11 | Ⓐ Ⓑ Ⓒ Ⓓ | 18 | Ⓐ Ⓑ Ⓒ Ⓓ | 25 Ⓐ Ⓑ Ⓒ Ⓓ |
| 5 | Ⓐ Ⓑ Ⓒ Ⓓ | 12 | Ⓐ Ⓑ Ⓒ Ⓓ | 19 | Ⓐ Ⓑ Ⓒ Ⓓ | 26 Ⓐ Ⓑ Ⓒ Ⓓ |
| 6 | Ⓐ Ⓑ Ⓒ Ⓓ | 13 | Ⓐ Ⓑ Ⓒ Ⓓ | 20 | Ⓐ Ⓑ Ⓒ Ⓓ | 27 Ⓐ Ⓑ Ⓒ Ⓓ |
| 7 | Ⓐ Ⓑ Ⓒ Ⓓ | 14 | Ⓐ Ⓑ Ⓒ Ⓓ | 21 | Ⓐ Ⓑ Ⓒ Ⓓ | |

28

29

30

31

# PSAT Mathematics Reference Sheet

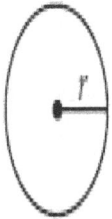 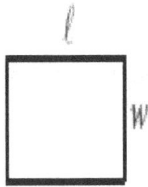 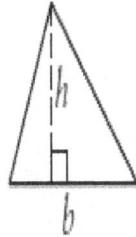 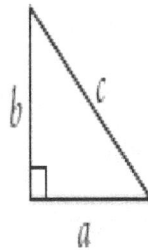 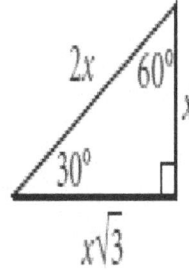 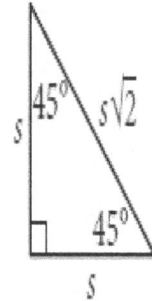

$A = \pi r^2$   $A = \ell w$   $A = \dfrac{1}{2}bh$   $c^2 = a^2 + b^2$   Special Right Triangles

$C = 2\pi r$

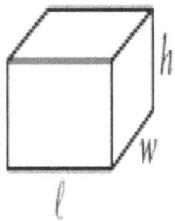   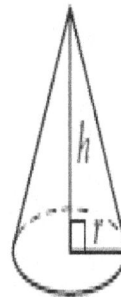 

$V = \ell wh$   $V = \pi r^2 h$   $V = \dfrac{4}{3}\pi r^3$   $V = \dfrac{1}{3}\pi r^2 h$   $V = \dfrac{1}{3}\ell wh$

The number of degrees of arc in a circle is 360.

The number of radians of arc in a circle is $2\pi$.

The sum of the measures in degrees of the angles of a triangle is 180.

# Sample of Grid-ins Answers

Write answers
in the boxes

Answer: 3.72

Answer: $\frac{6}{7}$

Grid in results

# PSAT Math Practice Test 1

## Section 1

❖ **17 Questions.**

❖ **Total time for this test: 25 Minutes**.

❖ **You may NOT use a calculator on this Section.**

**Administered** *Month Year*

# PSAT Subject Test – Mathematics

1) If $a$, $b$ and $c$ are positive integers and $3a = 7b = 2c$, then the value of $3a + 7b + 4c$ is how many times the value of $a$?

    A. 12                 C. 12.5

    B. 16                 D. 14

2) If $f(x^2) = 5x + 3$, for all positive value of $x$, what is the value of $f(121)$?

    A. $-58$            C. 56

    B. 58              D. $-56$

3) If $a$ and $b$ are solutions of the following equation, which of the following is the ratio $\frac{a}{b}$? $(a > b)$

$$2x^2 + 12x + 13 = 4x + 23$$

    A. $\frac{1}{5}$          C. $-\frac{1}{5}$

    B. $-5$          D. 5

4) If $x \neq -7$ and $x \neq 6$, which of the following is equivalent to $\frac{1}{\frac{1}{x-3}+\frac{1}{x+8}}$?

    A. $\frac{(x-3)(x+8)}{(x-3)+(x+8)}$       C. $\frac{(x+8)(x-3)}{(x+8)-(x+3)}$

    B. $\frac{(x+8)+(x-3)}{(x+8)(x-3)}$       D. $\frac{(x+8)+(x-3)}{(x+8)-(x-3)}$

5) A line in the $xy$-plane passes through origin and has a slope of $\frac{1}{3}$. Which of the following points lies on the line?

    A. (3,6)          C. (9, 3)

    B. (6,3)          D. (3,9)

6) Which of the following is the solution of the following inequality?

$$3x + 6.5 > 11x - 2.5 - 3.5x$$

A. $x < 2$             C. $x \le 5$

B. $x > 2$             D. $x \ge 5$

| Gender | Under 45 | 45 or older | total |
|--------|----------|-------------|-------|
| Male | 12 | 14 | 26 |
| Female | 16 | 8 | 24 |
| Total | 28 | 22 | 50 |

7) The table above shows the distribution of age and gender for 40 employees in a company. If one employee is selected at random, what is the probability that the employee selected be either a female under age 45 or a male age 45 or older?

A. $\frac{2}{5}$             C. $\frac{3}{25}$

B. $\frac{3}{5}$             D. $\frac{4}{25}$

8) If a parabola with equation $y = ax^2 + 3x + 15$, where $a$ is constant, passes through point $(2, 5)$, what is the value of $a^2$?

A. $-2$             C. $-16$

B. $2$             D. $16$

9) John works for an electric company. He receives a monthly salary of $4,100 plus 12% of all his monthly sales as bonus. If $x$ is the number of all John's sales per month, which of the following represents John's monthly revenue in dollars?

A. $0.12x$                           C. $0.12x + 4,100$

B. $0.88x - 4,100$                   D. $0.88x + 4,100$

10) What is the value of $f(2)$ for the following function $f$?

$$f(x) = x^4 - 8x$$

A. 1                                 C. 4

B. 0                                 D. 8

11) John buys a pepper plant that is 9 inches tall. With regular watering the plant grows 4 inches a year. Writing John's plant's height as a function of time, what does the $y$ −intercept represent?

A. The $y$ −intercept represents the rate of grows of the plant which is 9 in.

B. The $y$ −intercept represents the starting height of 9 in.

C. The $y$ −intercept represents the rate of growth of plant which is 4 in. per year.

D. There is no $y$ −intercept.

12) What is the length of AB in the following figure if AE = 5, CD = 3 and AC = 16?

A. 32

B. 8

C. 12

D. 10

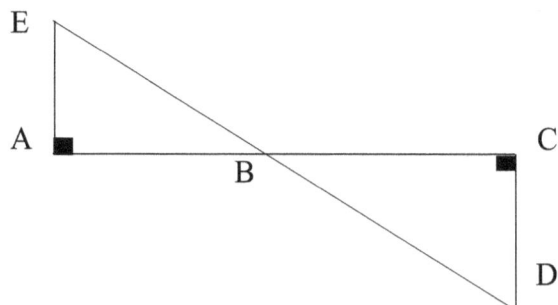

13) What is the solution of the following system of equations?

$$\begin{cases} \dfrac{-x}{8} + \dfrac{y}{4} = 3 \\ \dfrac{-5}{6}x + 2y = 24 \end{cases}$$

A. $x = 2, y = 12$

C. $x = 1, y = 10$

B. $x = 0\ y = 12$

D. $x = 10, y = 1$

## Grid-ins Questions

**Questions 14–17 are grid-ins questions. Solve the problems and enter your answers in the grid on the answer sheet as shown below.**

14) If $x \neq 0$, what is the value of $\dfrac{(6(x)(y^2))^4}{(4xy^2)^4}$?

15) In the following equation, what is the value of $y - 6x$?

$$\frac{y}{9} = x - \frac{x}{3} + 6$$

16) What is the value of $x$ in the following equation?

$$\frac{x^2 - 25}{x + 5} + 4(x + 2) = 20$$

17) The length of a rectangle is 3 meters greater than 4 times its width. The perimeter of the rectangle is 56 meters. What is the area of the rectangle in meters?

## STOP

**This is the End of this Section. You may check your work on this section if you still have time.**

# PSAT Math Practice Test 1

## Section 2

❖ **31 Questions.**

❖ **Total time for this test: 45 Minutes**.

❖ **You may use a calculator on this Section.**

**Administered** *Month Year*

1) If $y = nx + 7$, where $n$ is a constant, and when $x = 6$, $y = 14$, what is the value of $y$ when $x = 6$?

   A. 36                                C. 16

   B. 15                                D. 13

2) If $8 + 4x$ is 10 more than 14, what is the value of $8x$?

   A. 16                                C. 64

   B. 32                                D. 128

3) If a gas tank can hold 40 gallons, how many gallons does it contain when it is $\frac{5}{8}$ full?

   A. 40                                C. 25

   B. 50                                D. 64

4) In the $xy$-plane, the point $(2, 5)$ and $(1, 4)$ are online A. Which of the following equations of lines is parallel to line A?

   A. $y = 3x$                          C. $y = 2x$

   B. $y = \frac{x}{2}$                 D. $y = x$

5) A football team won exactly 40% of the games it played during last session. Which of the following could be the total number of games the team played last season?

   A. 43                                C. 48

   B. 45                                D. 34

6) The capacity of a red box is 25% bigger than the capacity of a blue box. If the red box can hold 50 equal sized books, how many of the same books can the blue box hold?

A. 15　　　　　　　　　　　　　　　　C. 25

B. 35　　　　　　　　　　　　　　　　D. 40

7) The sum of four different negative integers is $-46$. If the smallest of these integers is $-13$, what is the largest possible value of one of the other three integers?

A. $-11$　　　　　　　　　　　　　　C. $-10$

B. $-8$　　　　　　　　　　　　　　　D. $-9$

8) If $x$ is greater than 2 and less than 3, which of the following is true?

A. $x < \sqrt{x^2 + 2} < \sqrt{x^2} + 2$　　　　　　C. $\sqrt{x^2 + 2} < x < \sqrt{x^2} + 2$

B. $x < \sqrt{x^2} + 2 < \sqrt{x^2 + 2}$　　　　　　D. $\sqrt{x^2} + 2 < \sqrt{x^2 + 2} < x$

9) The ratio of boys and girls in a class is 4:5. If there are 72 students in the class, how many more boys should be enrolled to make the ratio 1:1?

A. 40　　　　　　　　　　　　　　　　C. 8

B. 12　　　　　　　　　　　　　　　　D. 32

10) If $f(x) = 2x + 3(x + 4) + 5$ then $f(2x) =$?

A. $10x + 17$　　　　　　　　　　　　C. $12x + 17$

B. $10x - 17$　　　　　　　　　　　　D. $12x - 17$

**Questions 11, 12 and 13 are based on the following data**

### Types of air pollutions in 10 cities of a country

| Type of Pollution | Number of Cities | | | | | | | | | |
|---|---|---|---|---|---|---|---|---|---|---|
| A | | | | | | | | | | |
| B | | | | | | | | | | |
| C | | | | | | | | | | |
| D | | | | | | | | | | |
| E | | | | | | | | | | |
| | 1 | 2 | 3 | 4 | 5 | 6 | 7 | 8 | 9 | 10 |

11) If $a$ is the mean (average) of the number of cities in each pollution type category, $b$ is the mode, and $c$ is the median of the number of cities in each pollution type category, then which of the following must be true?

A. $b < a < c$  

C. $a = c$

B. $c < a < b$  

D. $b < c = a$

12) How many cities should be added to type of pollutions C until the ratio of cities in type of pollution C to cities in type of pollution A will be 0.75?

A. 8  

C. 6

B. 4  

D. 2

13) What percent of cities are in the type of pollution B, C, and E respectively?

A. 60%, 60%, 20%  

C. 1.50%, 1.20%, 1.30%

B. 1.50%, 1.20%, 1.60%  

D. 50%, 20%, 60%

14) In the following right triangle, if the sides AB and BC become twice longer, what will be the ratio of the perimeter of the triangle to its area?

A. $\frac{1}{2}$

B. 3

C. $\frac{1}{3}$

D. 2

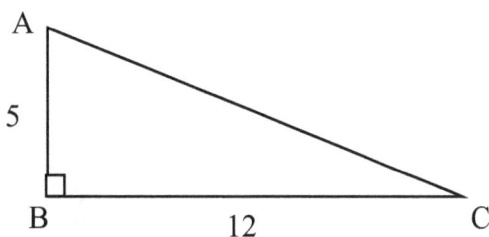

15) What is the ratio of the minimum value to the maximum value of the following function?

$$0 \leq x \leq 3$$

$$f(x) = -4x + 5$$

A. $\frac{9}{5}$

C. $-\frac{5}{9}$

B. $-\frac{7}{5}$

D. $\frac{5}{7}$

16) If $\frac{a-b}{b} = \frac{3}{8}$, then which of the following must be true?

A. $\frac{a}{b} = \frac{15}{8}$

C. $\frac{a}{b} = \frac{2}{11}$

B. $\frac{a}{b} = \frac{11}{8}$

D. $\frac{a}{b} = \frac{8}{11}$

**Questions 17 to 19 are based on the following data.**

The result of a research shows the number of men and women in four cities of a country.

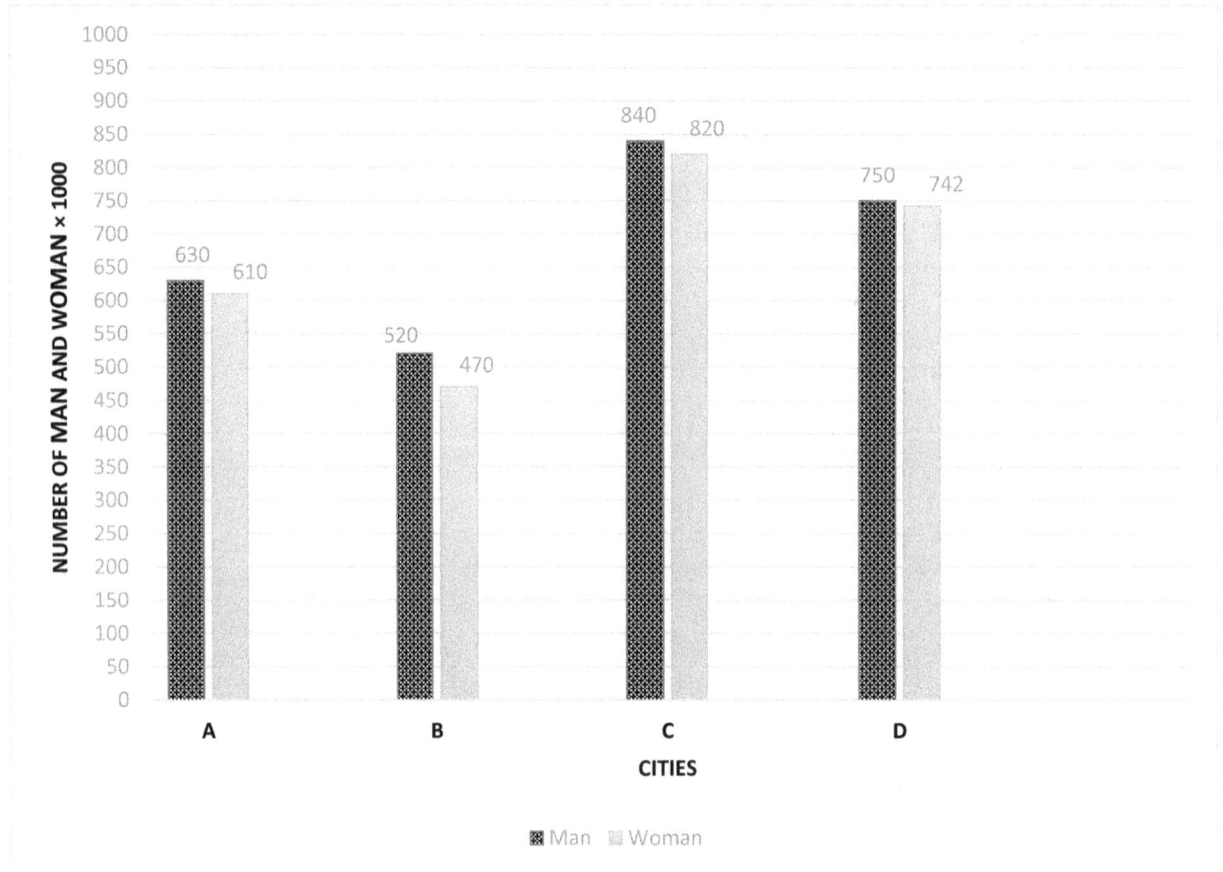

17) What's the ratio of percentage of men in city C to percentage of men in city B?

A. 0.104

C. 0.96

B. 9.633

D. 96.33

18) What's the minimum ratio of woman to man in the four cities?

A. 0.892

C. 0.904

B. 0.989

D. 0.976

19) How many women should be added to city D until the ratio of women to men will be 1.8?

    A. 338                C. 144

    B. 608                D. 308

20) In the rectangle below if $y > 6$ cm and the area of rectangle is 42 cm$^2$ and the perimeter of the rectangle is 26 cm, what is the value of $x$ and $y$ respectively?

    A. 7, 5

    B. 6, 13

    C. 6, 7

    D. 7, 13

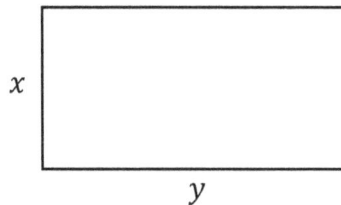

21) If a car has 70-liter petrol and after one hour driving the car use 3-liter petrol, how much petrol will remain after $x$-hours driving?

    A. $70x - 3$           C. $70 - 3x$

    B. $70x + 3$           D. $70 + 3x$

22) In the triangle below, if the measure of angle $A$ is 37 degrees, then what is the value of $y$? (figure is NOT drawn to scale)

    A. 55

    B. 50

    C. 75

    D. 70

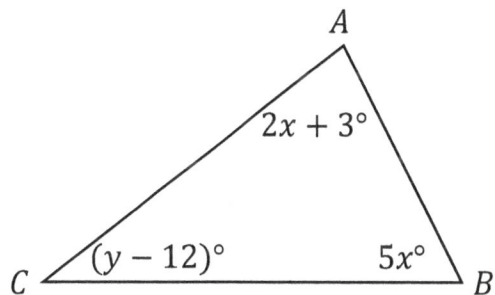

23) The following graph shows the mark of six students in mathematics. What is the mean (average) of the marks?

A. 13.5

B. 14

C. 14.5

D. 15

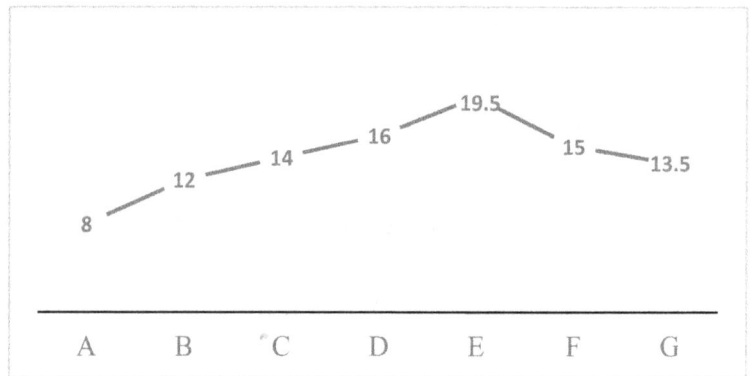

24) Which of the following values for $x$ and $y$ satisfy the following system of equations?

$$\begin{cases} x + 2y = 3 \\ -4x - 3y = -7 \end{cases}$$

A. $x = -1,\ y = 0$

B. $x = 0,\ y = 1$

C. $x = 1,\ y = -1$

D. $x = 1,\ y = 1$

25) Given the right triangle ABC bellow, cos ($\beta$) is equal to?

A. $\dfrac{c}{a}$

B. $\dfrac{a}{\sqrt{a^2+b^2}}$

C. $\dfrac{\sqrt{a^2+b^2}}{ab}$

D. $\dfrac{b}{\sqrt{a^2+b^2}}$

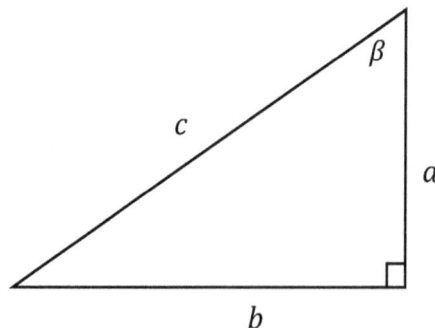

26) Solve the following inequality.

$$\left| \frac{x}{4} - x + 2 + 7 \right| < 3$$

   A. $-8 < x < 14$            C. $8 < x < 16$

   B. $-14 < x < 8$           D. $-16 < x < -8$

27) If $x$ is directly proportional to the square of $y$, and $y = 4$ when $x = 64$, then

when $x = 196$ $y = ?$

   A. $\frac{1}{9}$                 C. 7

                      D. 14

   B. 9

## Grid-ins Questions

**Questions 28–31 are grid-ins questions. Solve the problems and enter your answers in the grid on the answer sheet as shown below.**

28) $f(x) = ax^2 + bx + c$ is a quadratic function where $a$, $b$ and $c$ are constant, the value of $x$ of the point of intersection of this quadratic function and linear function $g(x) = 4x - 6$ is 3. The vertex of $f(x)$ is at $(-3, 0)$. What is the product of $a$, $b$ and $c$?

29) If $4x + 8y = \frac{-4y^2 + 12}{x}$, what is the value of $(x + y)^2$? $(x \neq 0)$

30) A ladder leans against a wall forming a 60° angle between the ground and the ladder. If the bottom of the ladder is 18 feet away from the wall, how many feet is the ladder?

31) The volume of cube A is $\frac{1}{2}$ of its surface area. What is the length of an edge of cube A?

## STOP

**This is the End of this Section. You may check your work on this section if you still have time.**

# PSAT Math Practice Test 2

# Section 1

❖ **17 Questions.**

❖ **Total time for this test: 25 Minutes**.

❖ **You may NOT use a calculator on this Section.**

**Administered** *Month Year*

## PSAT Subject Test – Mathematics

1) If $xp + 4yq = 15$ and $xp + 3yq = 9$, what is the value of $yq$?

   A. 10                                      C. 6

   B. 4                                         D. 8

2) If $x^2 + 4$ and $x^2 - 4$ are two factors of the polynomial $3x^4 + n$ and $n$ is a constant, what is the value of $n$?

   A. $-48$                                 C. 42

   B. $-24$                                 D. 32

3) If $5x - 4 = 9.5$, what is the value of $2x + 3$?

   A. 8.4                                    C. 10.4

   B. 9.4                                    D. 6.4

4) If the function $f$ is defined by $f(x) = x^2 + 3x - 7$, which of the following is equivalent to $f(4t^2)$?

   A. $16t^4 + 10t^2 - 7$                 C. $12t^4 + 7t^2 - 12t$

   B. $16t^4 + 12t^2 - 7$                 D. $16t^4 + 12t^2 + 7$

5) The circle graph below shows all Mr. Green's expenses for last month. If he spent \$550 on his car, how much did he spend for his rent?

   A. \$448

   B. \$884

   C. \$848

   D. \$484

Mr. Green's monthly expenses

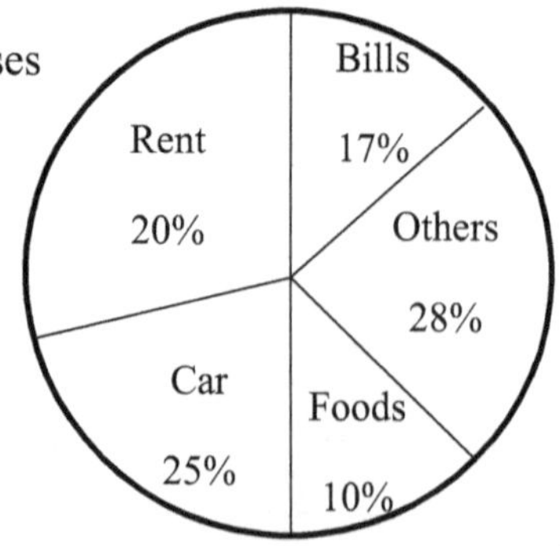

Rent 20%, Bills 17%, Others 28%, Foods 10%, Car 25%

$$0.ABC \qquad 0.0D$$

6) The letters represent two decimals listed above. One of the decimals is equivalent to $\frac{7}{8}$ and the other is equivalent to $\frac{1}{50}$. What is the product of C and D?

A. 2

C. 10

B. 5

D. 15

7) The radius of circle A is four times the radius of circle B. If the circumference of circle A is $16\pi$, what is the area of circle B?

A. $8\pi$

C. $4\pi$

B. $2\pi$

D. $16\pi$

8) In the diagram below, circle A represents the set of all even numbers, circle B represents the set of all negative numbers, and circle C represents the set of all multiples of 6. Which number could be replaced with $y$?

A. 0

B. 12

C. −18

D. −21

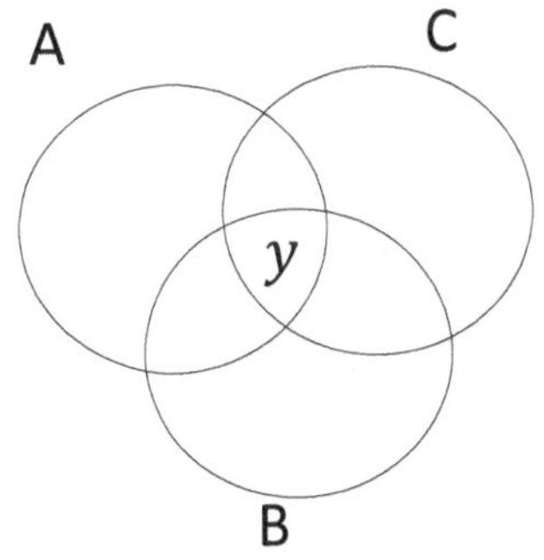

9) There are only red and blue cards in a box. The probability of choosing a red card in the box at random is one-third. If there are 144 blue cards, how many cards are in the box?

A. 162

C. 326

B. 362

D. 216

10) Both $(x = -2)$ and $(x = 1)$ are solutions for which of the following equations?

  I. $x^2 - 3x + 5 = 0$    III. $2x^2 + 2x - 4 = 0$

  II. $3x^2 - 3x = 6$

  A. I only    C. II and III

  B. I and III    D. I, II and III

11) In a certain bookshelf of a library, there are 50 biology books, 68 history books, and 82 language books. What is the ratio of the number of biology books to the total number of books in this bookshelf?

  A. $\dfrac{1}{6}$    C. $\dfrac{1}{5}$

  B. $\dfrac{1}{4}$    D. $\dfrac{3}{4}$

12) The following table represents the value of $x$ and function $f(x)$. Which of the following could be the equation of the function $f(x)$?

  A. $f(x) = x^2 + 2$    C. $f(x) = \sqrt{x + 3}$

  B. $f(x) = x^2 - 2$    D. $f(x) = \sqrt{x} + 3$

| $x$ | $f(x)$ |
| --- | --- |
| 1 | 4 |
| 4 | 5 |
| 9 | 6 |
| 16 | 7 |

13) In the figure below, what is the value of $x$?

  A. 52

  B. 63

  C. 65

  D. 117

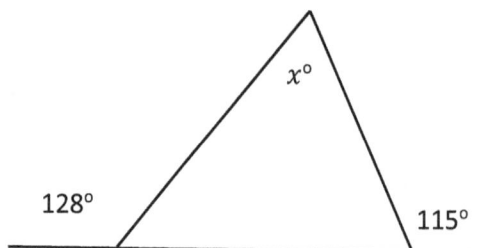

## Grid-ins Questions

**Questions 14–17 are grid-ins questions. Solve the problems and enter your answers in the grid on the answer sheet as shown below.**

14) If 16% of $x$ is 48 and $\frac{1}{5}$ of $y$ is 22, what is the value of $x - y$?

15) Michelle and Alec can finish a job together in 30 minutes. If Michelle can do the job by herself in 3 hours, how many minutes does it take Alec to finish the job?

16) Angle $a$ is 630 degrees and can be written $x\pi$ in radian. What is the value of $x$?

17) In the following figure, point O is the center of the circle and the equilateral triangle has perimeter 36. What is the circumference of the circle? ($\pi = 3$)

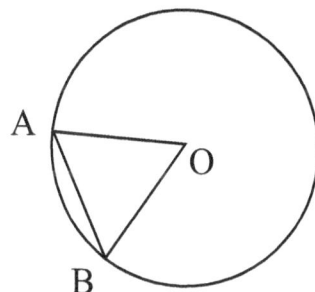

## STOP

**This is the End of this Section. You may check your work on this section if you still have time.**

# PSAT Math Practice Test 2

# Section 2

❖ **31 Questions.**

❖ **Total time for this test: 45 Minutes**.

❖ **You may use a calculator on this Section.**

**Administered** *Month Year*

1) If $(4^a)^b = 256$, then what is the value of $ab$?

   A. 2                            C. 4

   B. 6                            D. 8

2) What is the sum of $\sqrt{x-9}$ and $\sqrt{x} - 9$ when $\sqrt{x} = 5$ ?

   A. −4                         C. 0

   B. −1                         D. 4

3) What is the average (arithmetic mean) of all integers from 15 to 21?

   A. 17                         C. 18

   B. 18.5                      D. 19.5

4) What is the value of $\frac{7a-1}{6}$, if $-3a + 5a + 7a = 63$ ?

   A. 7.5                       C. 8

   B. 7                          D. 8.5

5) What is the value of $|-16 - 7| - |-12 + 3|$?

   A. −14                       C. 12

   B. 14                       D. −12

6) A container holds 1.6 gallons of water when it is $\frac{8}{35}$ full. How many gallons of water does the container hold when it's full?

   A. 8                          C. 12

   B. 7                         D. 15

7) The table represents different values of function $g(x)$. What is the value

of $4g(-2) - 2g(4)$?

A. $-20$

B. $-12$

C. 12

D. 20

| $x$ | $-2$ | $-1$ | 0 | 2 | 3 | 4 |
|---|---|---|---|---|---|---|
| $g(x)$ | 3 | 2 | 1 | 0 | $-2$ | $-4$ |

8) If $a$ is an odd integer divisible by 7. Which of the following must be divisible by 5?

A. $a - 4$

B. $a + 4$

C. $4a$

D. $5a - 5$

9) If $(x - 3)^2 = 9$ which of the following could be the value of $(x - 2)(x - 5)$?

A. 3

B. 4

C. $-4$

D. $-3$

10) On the following figure, what is the area of the quadrilateral ABCD?

A. 12

B. 32

C. 24

D. 42

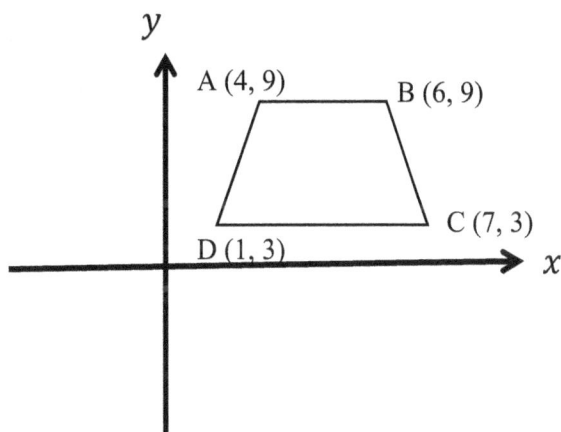

**Questions 11 to 13 are base**

Number of clothes sold in a clothing store

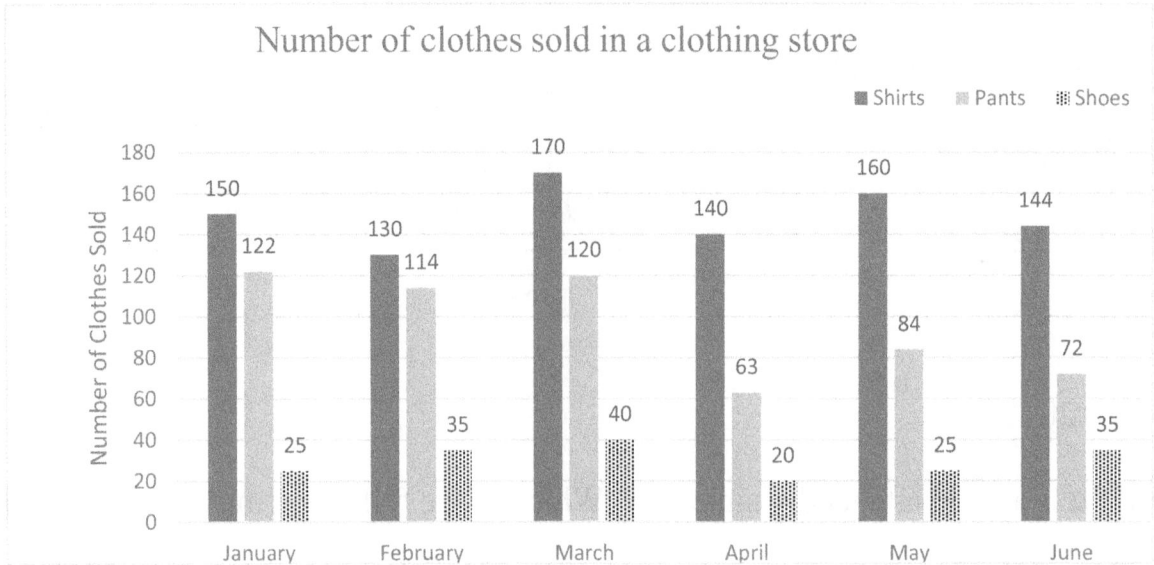

11) Between which two of the months shown was there a twenty-five percent increase in the number of pants sold?

A. January and February

C. March and April

B. February and March

D. April and May

12) During the six-month period shown, what is the mean number of shirts and median number of shoes per month?

A. 40, 147.5

C. 149, 30

B. 139, 30

D. 25, 145

13) How many shoes need to be added in February until the ratio of number of pants to number of shoes in February equals to three-twelfth of this ratio in March?

A. 107

C. 11

B. 117

D. 17

14) What is the $x$-intercept of the line with equation $5x - 3y = 8$?

A. 3

B. $-5$

C. $\frac{8}{5}$

D. $\frac{3}{5}$

15) The base of a right triangle is 10 feet, and the interior angles are 45-45-90. What is its area?

A. 50 square feet

B. 10 square feet

C. 40 square feet

D. 100 square feet

16) In 1999, the average worker's income increased $3,000 per year starting from $19,000 annual salary. Which equation represents income greater than average? (I = income, x = number of years after 1999)

A. $I > 3,000\, x + 19,000$

B. $I > -3,000\, x + 19,000$

C. $I < -3,000\, x + 19,000$

D. $I < 3,000\, x - 19,000$

17) The Jackson Library is ordering some bookshelves. If $x$ is the number of bookshelves the library wants to order, which each cost $200 and there is a one-time delivery charge of $500, which of the following represents the total cost, in dollar, per bookshelf?

A. $200x + 500$

B. $200 + 500x$

C. $\frac{200x+500}{200}$

D. $\frac{200x+500}{x}$

A library has 500 books that include Mathematics, Physics, Chemistry, English, and History.

Use following graph to answer questions 18 to 20.

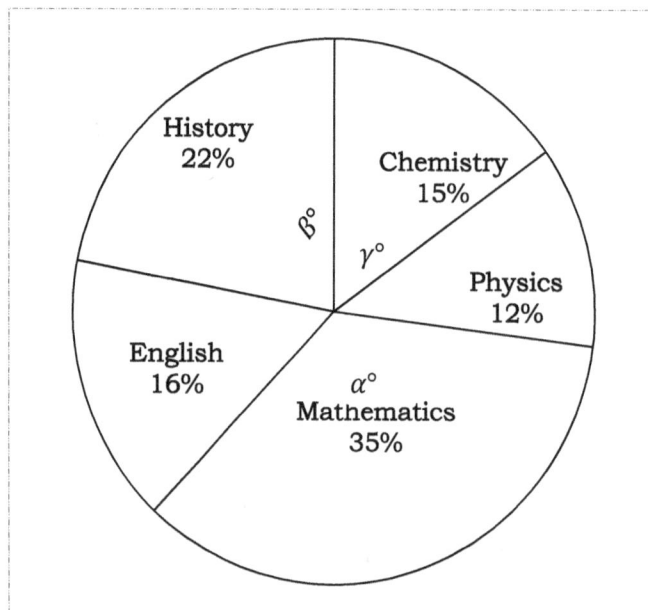

18) What is the product of the number of Mathematics and number of Chemistry books?

A. 13,125

C. 23,125

B. 23,351

D. 13,351

19) The librarians decided to move some of the books in the Chemistry section to Mathematics section. How many books are in the Chemistry section if now $\gamma = \frac{3}{7}\alpha$?

A. 60

C. 117

B. 210

D. 75

20) What are the values of angle $\gamma$ and $\beta$ respectively?

  A. 60°, 120°                    C. 52°, 65.5°

  B. 75°, 105°                    D. 54°, 79.2°

21) In the following figure, point Q lies online n, what is the value of $y$ if $x = 30$?

  A. 15

  B. 30

  C. 25

  D. 20

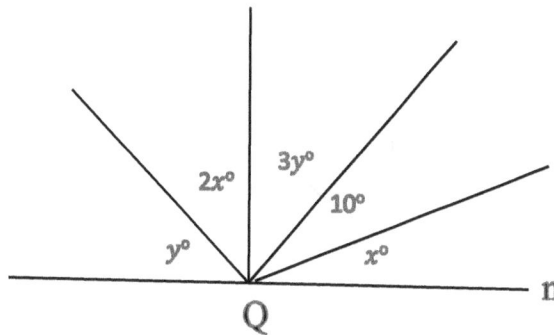

22) In the following figure, AB is the diameter of the circle. What is the circumference of the circle?

  A. $5\pi$

  B. $7.5\pi$

  C. $2.5\pi$

  D. $10\pi$

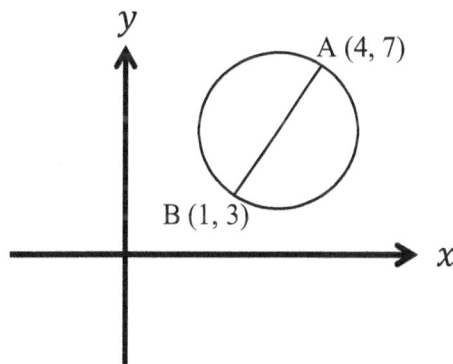

23) What is the smallest integer whose square root is greater than 7?

  A. 16                    C. 48

  B. 9                     D. 64

24) If the area of trapezoid is 104 cm, what is the perimeter of the trapezoid?

A. 40 cm

B. 24 cm

C. 44 cm

D. 48 cm

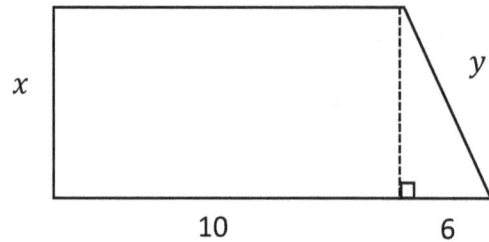

25) What is the solution of the following inequality?

$$|x - 3| \geq 12$$

A. $x \geq 15 \ \cup \ x \leq -9$

C. $x \geq 15$

B. $-9 \leq x \leq 15$

D. $x \leq -9$

26) If the area of the following rectangular ABCD is 120, and E is the midpoint of AB, what is the area of the shaded part?

A. 30

B. 60

C. 40

D. 80

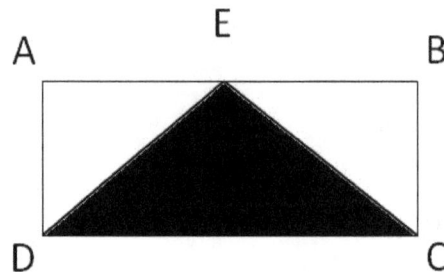

27) Which of the following is equivalent to $5 < -4x - 3 < 13$?

A. $-4 < x < -2$

C. $3 < x < 8$

B. $2 < x < 4$

D. $\frac{-3}{5} < x < \frac{1}{2}$

## Grid-ins Questions

**Questions 28–31 are grid-ins questions. Solve the problems and enter your answers in the grid on the answer sheet as shown below.**

28) In the following figure, ABCD is a rectangle. If $a = \sqrt{3}$, and $b = 4a$, find the area of the shaded region? (the shaded region is a trapezoid) (Round your answer to the nearest hundredths place)

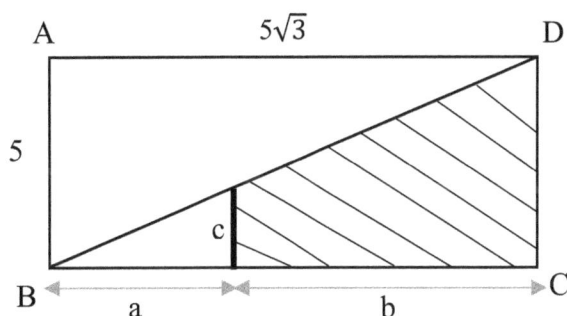

29) 9 liters of water are poured into an aquarium that's 18cm long, 5cm wide, and 70cm high. How many cm will the water level in the aquarium rise due to this added water? (1 liter of water $= 1,000 \; cm^3$)?

30) If $x \begin{bmatrix} 4 & 0 \\ 0 & 5 \end{bmatrix} = \begin{bmatrix} 3x + 2y - 3 & 0 \\ 0 & 8y - 9 \end{bmatrix}$, what is the product of $x$ and $y$?

31) If $cos\ A = \frac{\sqrt{2}}{2}$ in a right triangle and the angle A is an acute angle, then what is $sin\ A$? (Round your answer to the nearest hundredths place)

**STOP**

**This is the End of this Section. You may check your work on this section if you still have time.**

# Chapter 17 :

# Answers and Explanations

## Answer Key

❊ Now, it's time to review your results to see where you went wrong and what areas you need to improve!

### PSAT Math Practice Test 1

| **Section 1 – No Calculator** | | | | **Section 2 - Calculator** | | | | | |
|---|---|---|---|---|---|---|---|---|---|
| 1 | A | 14 | $\frac{81}{16}$ | 1 | D | 14 | A | 27 | C |
| 2 | B | 15 | 54 | 2 | B | 15 | B | 28 | $\frac{1}{4}$ |
| 3 | B | 16 | $\frac{17}{5}$ | 3 | C | 16 | B | 29 | 3 |
| 4 | A | 17 | 115 | 4 | D | 17 | D | 30 | 36 |
| 5 | C | | | 5 | B | 18 | C | 31 | 3 |
| 6 | A | | | 6 | D | 19 | B | | |
| 7 | B | | | 7 | C | 20 | C | | |
| 8 | D | | | 8 | A | 21 | C | | |
| 9 | C | | | 9 | C | 22 | D | | |
| 10 | B | | | 10 | A | 23 | B | | |
| 11 | B | | | 11 | C | 24 | D | | |
| 12 | D | | | 12 | B | 25 | B | | |
| 13 | B | | | 13 | D | 26 | C | | |

# Answers and Explanations

## PSAT Math Practice Test 2

| Section 1- No Calculator | | | | Section 2- Calculator | | | | | |
|---|---|---|---|---|---|---|---|---|---|
| 1 | C | **14** | 190 | 1 | C | **14** | C | **27** | A |
| 2 | A | **15** | 36 | 2 | C | **15** | A | **28** | 16.97 |
| 3 | A | **16** | 3.5 | 3 | C | **16** | A | **29** | 100 |
| 4 | B | **17** | 72 | 4 | C | **17** | D | **30** | 9 |
| 5 | D | | | 5 | B | **18** | A | **31** | $\frac{\sqrt{2}}{2}$ |
| 6 | C | | | 6 | B | **19** | D | | |
| 7 | C | | | 7 | D | **20** | D | | |
| 8 | C | | | 8 | D | **21** | D | | |
| 9 | D | | | 9 | B | **22** | A | | |
| 10 | C | | | 10 | C | **23** | D | | |
| 11 | B | | | 11 | D | **24** | C | | |
| 12 | D | | | 12 | C | **25** | A | | |
| 13 | B | | | 13 | B | **26** | B | | |

# Practice Tests 1:

# Section 1

**1) Answer: A.**

$$3a = 7b \rightarrow b = \frac{3a}{7} \quad \text{and } 3a = 2c \rightarrow c = \frac{3a}{2}$$

$$3a + 7b + 4c = 3a + \left(7 \times \frac{3a}{7}\right) + \left(4 \times \frac{3a}{2}\right) = 3a + 3a + 6a = 12a \; ; \text{The value of}$$

$3a + 7b + 2c$ is 12 times the value of $a$.

**2) Answer: B.**

$x^2 = 121 \rightarrow x = 11$ (positive value)  Or $x = -11$(negative value)

Since $x$ is positive, then:

$$f(121) = f(11^2) = 5(11) + 3 = 55 + 3 = 58$$

**3) Answer: B.**

$$2x^2 + 12x + 13 = 4x + 23 \rightarrow 2x^2 - 4x + 12x + 13 - 23 = 0 \rightarrow 2x^2 + 8x - 10 = 0$$

$$\rightarrow 2(x^2 + 4x - 5) = 0 \rightarrow \text{Divide both sides by 2. Then: } x^2 + 4x - 5 = 0,$$

Find the factors of the quadratic equation. $\rightarrow (x + 5)(x - 1) = 0 \rightarrow x = -5$ or $x = 1$

$a > b$, then: $a = -5$ and $b = 1 \rightarrow \frac{a}{b} = \frac{-5}{1} = -5$

**4) Answer: A.**

To rewrite $\frac{1}{\frac{1}{x-3}+\frac{1}{x+8}}$, first simplify $\frac{1}{x-3} + \frac{1}{x+8}$.

$$\frac{1}{x-3} + \frac{1}{x+8} = \frac{(x+8)}{(x-3)(x+8)} + \frac{(x-3)}{(x+8)(x-3)} = \frac{(x+8)+(x-3)}{(x+8)(x-3)}$$

Then: $\frac{1}{\frac{1}{x-3}+\frac{1}{x+8}} = \frac{1}{\frac{(x+8)+(x-3)}{(x+8)(x-3)}} = \frac{(x+8)(x-3)}{(x+8)+(x-3)}$. (Remember, $\frac{1}{\frac{1}{x}} = x$)

This result is equivalent to the expression in choice A.

**5) Answer: C.**

First, find the equation of the line. All lines through the origin are of the form $y = mx$,

so the equation is $y = \frac{1}{3}x$. Of the given choices, only choice C (9,3), satisfies this

equation: $y = \frac{1}{3}x \rightarrow 3 = \frac{1}{3}(9) = 3$

**6) Answer: A.**

$3x + 6.5 > 11x - 2.5 - 3.5x \rightarrow$ Combine like terms:

$3x + 6.5 > 7.5x - 2.5 \rightarrow$ Subtract $3x$ from both sides:

$6.5 > 4.5x - 2.5$

Add 2.5 both sides of the inequality.

$9 > 4.5x$, Divide both sides by 4.5: $\frac{9}{4.5} > x \rightarrow x < 2$

**7) Answer: B.**

Of the 40 employees, there are 16 females under age 45 and 14 males age 45 or older.

Therefore, the probability that the person selected will be either a female under age 45

or a male age 45 or older is: $\frac{16}{50} + \frac{14}{50} = \frac{30}{50} = \frac{3}{5}$

**8) Answer: D.**

Plug in the values of $x$ and $y$ of the point $(2, 5)$ in the equation of the parabola. Then:

$5 = a(2)^2 + 3(2) + 15 \rightarrow 5 = 4a + 6 + 15 \rightarrow 5 = 4a + 21$

$\rightarrow 4a = 5 - 21 = -27 \rightarrow a = \frac{-16}{4} = -4 \rightarrow a^2 = (-4)^2 = 16$

**9) Answer: C.**

$x$ is the number of all John's sales per month and 12% of it is:

$12\% \times x = 0.12x$

John's monthly revenue: $0.12x + 4,100$

**10) Answer: B.**

The output value is 0. Then: $x = 2$

$f(x) = x^4 - 8x \rightarrow f(2) = 2^4 - 8(2) = 16 - 16 = 0$

**11) Answer: B.**

To solve this problem, first recall the equation of a line: $y = mx + b$ ; Where, $m =$

*slope* and $y = y - intercept$

Remember that slope is the rate of change that occurs in a function and that the

$y -$intercept is the $y$ value corresponding to $x = 0$.

Since the height of John's plant is 9 inches tall when he gets it. Time (or $x$) is zero. The plant grows 4 inches per year. Therefore, the rate of change of the plant's height is 4. The $y$ −intercept represents the starting height of the plant which is 9 inches.

**12) Answer: D.**

Two triangles $\Delta BAE$ and $\Delta BCD$ are similar. Then:

$\frac{AE}{CD} = \frac{AB}{BC} \to \frac{5}{3} = \frac{x}{16-x} \to 5(16 - x) = 3x \to 5x + 3x = 5 \times 16 \to 8x = 80 \to x = 10$

**13) Answer: B.**

$\begin{cases} \frac{-x}{8} + \frac{y}{4} = 3 \\ \frac{-5}{6}x + 2y = 24 \end{cases}$ Multiply the top equation by $-8$. Then, $\begin{cases} x - 2y = -24 \\ \frac{-5}{6}x + 2y = 24 \end{cases}$ Add two equations.

$\frac{1}{6}x = 0 \to x = 0$ , plug in the value of x into the first equation $\to y = 12$

**14) Answer: $\frac{81}{16}$.**

First, simplify the numerator and the denominator.

$\frac{(6(x)(y^2))^4}{(4xy^2)^4} = \frac{1,296x^4y^8}{256x^4y^8}$

Remove $x^4y^8$ from both numerator and denominator. $\frac{1,296x^4y^8}{256x^4y^8} = \frac{1,256}{256} = \frac{81}{16}$

**15) Answer: 54.**

$\frac{y}{9} = x - \frac{x}{3} + 6$, Multiply both sides of the equation by 9. Then:

$9 \times \frac{y}{9} = 9 \times \left(x - \frac{x}{3} + 6\right) \to y = 9x - 3x + 54 \to y = 6x + 54$

Now, subtract $6x$ from both sides of the equation. Then, $y - 6x = 54$

**16) Answer: $\frac{17}{5}$.**

First, factorize the numerator and simplify.

$\frac{x^2-25}{x+5} + 4(x + 2) = 20 \to \frac{(x-5)(x+5)}{x+5} + 4x + 8 = 20$

Divide both sides of the fraction by $(x + 5)$. Then:

$x - 5 + 4x + 8 = 20 \to 5x + 3 = 20$

Subtract 3 from both sides of the equation. Then: $5x = 20 - 3 = 17 \to x = \frac{17}{5}$

**17) Answer: 115.**

Let $L$ be the length of the rectangular and $W$ be the with of the rectangular. Then, $L = 4W + 3$

The perimeter of the rectangle is 56 meters. Therefore:

$2L + 2W = 56 \Longrightarrow L + W = 28$

Replace the value of $L$ from the first equation into the second equation and solve for

$W: (4W + 3) + W = 28 \rightarrow 5W + 3 = 28 \rightarrow 5W = 25 \rightarrow W = 5$

The width of the rectangle is 5 meters, and its length is:

$L = 4W + 3 = 4(5) + 3 = 23$

The area of the rectangle is: length × width = $5 \times 23 = 115$

# Practice Tests 1:

# Section 2

**1) Answer: D.**

Substituting 6 for $x$ and 14 for $y$ in $y = nx + 7$ gives $14 = (n)(6) + 7$

which gives $n = 1$. Hence, $y = x + 7$. Therefore, when $x = 6$, the value of $y$ is:

$$y = 6 + 7 = 13$$

**2) Answer: B.**

The description $8 + 4x$ *is* 10 more than 14 can be written as the equation $8 + 4x = 10 + 14$, which is equivalent to $8 + 4x = 24$. Subtracting 8 from each side gives:

$4x = 16$.

Since $8x$ is 2 times $4x$, multiplying both sides of $4x = 16$ by 2 gives $8x = 32$

**3) Answer: C.**

$\frac{5}{8} \times 40 = \frac{200}{8} = 25$

**4) Answer: D.**

The slop of line A is: $m = \frac{y_2 - y_1}{x_2 - x_1} = \frac{5-4}{2-1} = 1$

Parallel lines have the same slope and only choice D ($y = x$) has slope of 1.

**5) Answer: B.**

Choices A, C and D are incorrect because 40% of each of the numbers is a non-whole number.

    A. 43, $40\%$ $of$ $43 = 0.40 \times 43 = 17.2$

    B. 45, $40\%$ $of$ $45 = 0.40 \times 45 = 18$

    C. 48, $40\%$ $of$ $48 = 0.40 \times 48 = 19.2$

    D. 34, $40\%$ $of$ $34 = 0.40 \times 34 = 13.6$

Only choice B gives a whole number.

**6) Answer: D.**

The capacity of a red box is 25% bigger than the capacity of a blue box and it can hold 50 books. Therefore, we want to find a number that 25% bigger than that number is

50. Let $x$ be that number. Then: $1.25 \times x = 50$, Divide both sides of the equation by 1.25. Then: $x = \frac{50}{1.25} = 40$

## 7) Answer: C.

The smallest number is $-13$. To find the largest possible value of one of the other three integers, we need to choose the smallest possible integers for three of them. Let $x$ be the largest number. Then: $-46 = (-13) + (-12) + (-11) + x \rightarrow -46 = -36 + x \rightarrow x = -46 + 36 = -10$

## 8) Answer: A.

Let $x$ be equal to 2.5, then: $x = 2.5$

$\sqrt{x^2 + 2} = \sqrt{2.5^2 + 2} = \sqrt{8.25} \approx 2.87$

$\sqrt{x^2} + 2 = \sqrt{2.5^2} + 2 = 2.5 + 2 = 4.5$

Then, option A is correct: $x < \sqrt{x^2 + 2} < \sqrt{x^2} + 2$

## 9) Answer: C.

The ratio of boy to girls is 4:5. Therefore, there are 4 boys out of 9 students. To find the answer, first divide the total number of students by 9, then multiply the result by 4.

$72 \div 9 = 8 \Rightarrow 4 \times 8 = 32$

There are 32 boys and 40 (72 – 32) girls. So, 8 more boys should be enrolled to make the ratio 1:1.

## 10) Answer: A.

If $f(x) = 2x + 3(x + 4) + 5$, then find $f(2x)$ by substituting $2x$ for every $x$ in the function. This gives: $f(2x) = 2(2x) + 3(2x + 4) + 5$,

It simplifies to: $f(2x) = 4x + 6x + 12 + 5 = 10x + 17$

## 11) Answer: C.

Let's find the mean (average), mode and median of the number of cities for each type of pollution.

Number of cities for each type of pollution: 6, 2, 5, 8, 9

average (mean) $= \frac{sum\ of\ terms}{number\ of\ terms} = \frac{6+2+5+8+9}{5} = \frac{30}{5} = 6$

Median is the number in the middle. To find median, first list numbers in order from smallest to largest: 2, 5, 6, 8, 9. Median of the data is 6.

Mode is the number which appears most often in a set of numbers. Therefore, there is no mode in the set of numbers. Median = Mean, then, $a=c$

## 12) Answer: B.

Let the number of cities should be added to type of pollutions C be $x$. Then: $\frac{x+2}{8} = 0.75 \rightarrow x + 2 = 8 \times 0.75 \rightarrow x + 2 = 6 \rightarrow x = 4$

## 13) Answer: D.

Percent of cities in the type of pollution B: $\frac{5}{10} \times 100 = 50\%$

Percent of cities in the type of pollution C: $\frac{2}{10} \times 100 = 20\%$

Percent of cities in the type of pollution E: $\frac{6}{10} \times 100 = 60\%$

## 14) Answer: A.

$AB = 5$, and $BC = 12$; $AC = \sqrt{5^2 + 12^2} = \sqrt{25 + 144} = \sqrt{169} = 13$

Perimeter $= 5 + 12 + 13 = 30$

Area $= \frac{5 \times 12}{2} = 5 \times 6 = 30$

In this case, the ratio of the perimeter of the triangle to its area is: $\frac{30}{30} = 1$

If the sides AB and BC become twice longer, then: $AB = 10$ And $BC = 24$

$AC = \sqrt{10^2 + 24^2} = \sqrt{100 + 576} = \sqrt{676} = 26$

Perimeter $= 26 + 24 + 10 = 60$

Area $= \frac{10 \times 24}{2} = 10 \times 12 = 120$

In this case the ratio of the perimeter of the triangle to its area is: $\frac{60}{120} = \frac{1}{2}$

## 15) Answer: B.

Since $f(x)$ is linear function with a negative slop, then when $x = 0$, $f(x)$ is maximum and when $x = 3$, $f(x)$ is minimum. Then the ratio of the minimum value to the maximum value of the function is: $\frac{f(3)}{f(0)} = \frac{-4(3)+5}{-4(0)+5} = \frac{-7}{5}$

# PSAT Subject Test – Mathematics

**16) Answer: B.**

The equation $\frac{a-b}{b} = \frac{3}{8}$ can be rewritten as $\frac{a}{b} - \frac{b}{b} = \frac{3}{8}$, from which it follows that $\frac{a}{b} - 1 = \frac{3}{8}$, or $\frac{a}{b} = \frac{3}{8} + 1 = \frac{11}{8}$.

**17) Answer: D.**

Percentage of men in city C $= \frac{840}{1,660} \times 100 = 50.60\%$

Percentage of men in city B $= \frac{520}{990} \times 100 = 52.53\%$

Percentage of men in city C to percentage of men in city B: $\frac{50.60}{52.53} = 96.33$

**18) Answer: C.**

Ratio of women to men in city A: $\frac{610}{630} = 0.968$

Ratio of women to men in city B: $\frac{470}{520} = 0.904$

Ratio of women to men in city C: $\frac{820}{840} = 0.976$

Ratio of women to men in city D: $\frac{742}{750} = 0.989$

**19) Answer: B.**

Let the number of women should be added to city D be x, then:

$\frac{742 + x}{750} = 1.8 \rightarrow 742 + x = 750 \times 1.8 = 1,350 \rightarrow x = 608$

**20) Answer: C.**

The perimeter of the rectangle is: $2x + 2y = 26 \rightarrow x + y = 13 \rightarrow x = 13 - y$

The area of the rectangle is: $x \times y = 42 \rightarrow (13 - y)(y) = 42 \rightarrow y^2 - 13y + 42 = 0$

Solve the quadratic equation by factoring method. $(y - 6)(y - 7) = 0$

$y = 6$ (Unacceptable, because $y$ must be greater than 6)

or $y = 7 \rightarrow x \times y = 42 \rightarrow x \times 7 = 42 \rightarrow x = 6$

**21) Answer: C.**

The amount of petrol consumed after $x$ hours is: $3 \times x = 3x$

Petrol remaining after $x$ hours driving: $70 - 3x$

**22) Answer: D.**

In the figure angle $A$ is labeled $(2x + 3)$ and it measures 37. Thus, $2x + 3 = 37$ and $2x = 34$ or $x = 17$.

That means that angle $B$, which is labeled $(5x)$, must measure $5 \times 17 = 85$.

Since the three angles of a triangle must add up to 180, $37 + 85 + y - 12 = 180$, then: $y + 110 = 180 \rightarrow y = 180 - 110 = 70$

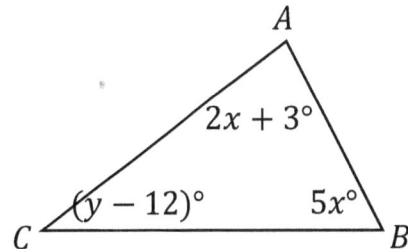

**23) Answer: B.**

$$average \ (mean) = \frac{sum \ of \ terms}{number \ of \ terms} = \frac{8+12+14+16+19.5+15+13.5}{7} = 14$$

**24) Answer: D.**

$$\begin{cases} x + 2y = 3 \\ -4x - 3y = -7 \end{cases} \rightarrow \text{Multiply the top equation by 4 then,}$$

$$\begin{cases} 4x + 8y = 12 \\ -4x - 3y = -7 \end{cases} \rightarrow \text{Add two equations.}$$

$5y = 5 \rightarrow y = 1$ , plug in the value of $y$ into the first equation:

$x + 2y = 3 \rightarrow x + 2(1) = 3 \rightarrow x + 2 = 3$

Add $-2$ from both sides of the equation.

Then: $x + 2 - 2 = 3 - 2 \rightarrow x = 1$

**25) Answer: B.**

$$\sin \beta = \frac{Opposite \ side}{hypotenuse}$$

To find the hypotenuse, we need to use Pythagorean theorem.

$a^2 + b^2 = c^2 \rightarrow c = \sqrt{a^2 + b^2}$

$\cos (\beta) = \frac{a}{c} = \frac{a}{\sqrt{a^2 + b^2}}$

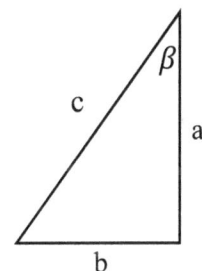

**26) Answer: C.**

$$\left| \frac{x}{4} - x + 2 + 7 \right| < 3 \rightarrow \left| -\frac{3}{4}x + 9 \right| < 3 \rightarrow -3 < -\frac{3}{4}x + 9 < 3$$

Subtract 9 from all sides of the inequality.

$\rightarrow -3 - 9 < -\frac{3}{4}x + 9 - 9 < 3 - 9 \rightarrow -12 < -\frac{3}{4}x < -6$

Multiply all sides by 4.

$\rightarrow 4 \times (-12) < 4 \times \left(-\frac{3x}{4}\right) < 4 \times (-6) \rightarrow -48 < -3x < -24$

Divide all sides by $-3$. (Remember that when you divide all sides of an inequality by a negative number, the inequality sing will be swapped. $<$ becomes $>$)

$\rightarrow \frac{-48}{-3} > \frac{-3x}{-3} > \frac{-24}{-3} \rightarrow 16 > x > 8 \rightarrow 8 < x < 16$

### 27) Answer: C.

$x$ is directly proportional to the square of $y$. Then: $x = cy^2$

$64 = c(4)^2 \rightarrow 64 = 16c \rightarrow c = \frac{64}{16} = 4$

The relationship between $x$ and $y$ is: $x = 4y^2$

$x = 196 \rightarrow 196 = 4y^2 \rightarrow y^2 = \frac{196}{4} = 49 \rightarrow y = 7$

### 28) Answer: $\frac{1}{4}$.

The intersection of two functions is the point with 3 for $x$. Then:

$f(3) = g(3)$ and $g(3) = \left(4 \times (3)\right) - 6 = 12 - 6 = 6$

Then, $f(3) = 6 \rightarrow a(3)^2 + b(3) + c = 6 \rightarrow 9a + 3b + c = 6$  (i)

The value of $x$ in the vertex of the parabola is:

$x = -\frac{b}{2a} \rightarrow -3 = -\frac{b}{2a} \rightarrow b = 6a \,(ii)$

In the point $(-3, 0)$, $f(-3) = 0 \rightarrow a(-3)^2 + b(-3) + c = 0$

$\rightarrow 9a - 3b + c = 0 (iii)$

Using the first two equation: $\begin{cases} 9a + 3b + c = 6 \\ 9a - 3b + c = 0 \end{cases}$

Equation 1 minus equation 3 is: (i)$-$(iii) $\rightarrow 6b = 6 \rightarrow b = 1$ (iv)

Plug in the value of $b$ in the second equation: $b = 6a \rightarrow a = \frac{b}{6} = \frac{1}{6} = \frac{1}{6}$

Plug in the values of a and b in the first equation. Then:

$9\left(\frac{1}{6}\right) + 3(1) + c = 6 \rightarrow \frac{3}{2} + 3 + c = 6 \rightarrow c = 3 - \frac{3}{2} \rightarrow c = \frac{3}{2}$

The product of $a$, $b$ and c $=\left(\frac{1}{6}\right) \times 1 \times \left(\frac{3}{2}\right) = \frac{1}{4}$

**29) Answer: 3.**

$4x + 8y = \frac{-4y^2+12}{x}$, Multiply both sides by $x$.

$x \times (4x + 8y) = x \times \left(\frac{-4y^2+12}{x}\right) \rightarrow 4x^2 + 8xy = -4y^2 + 12$

$\rightarrow 4x^2 + 8xy + 4y^2 = 12 \rightarrow 4 \times (x^2 + 2xy + y^2) = 12 \rightarrow x^2 + 2xy + y^2 = \frac{12}{4}$

$x^2 + 2xy + y^2 = (x + y)^2$, Then: $(x + y)^2 = 3$

**30) Answer: 36 feet.**

The relationship among all sides of special right triangle

$30°, 60°, 90°$ is provided in this triangle:

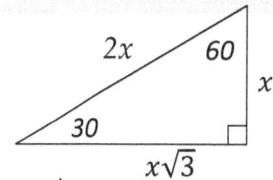

In this triangle, the opposite side of $30°$ angle is half of the

hypotenuse.

Draw the shape of this question.

The latter is the hypotenuse. Therefore, the latter is 36 feet.

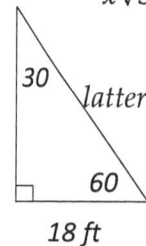

**31) Answer: 3.**

Let $x$ be the length of an edge of cube, then the volume of a cube is: $V = x^3$

The surface area of cube is: $SA = 6x^2$

The volume of cube A is $\frac{1}{2}$ of its surface area. Then, $x^3 = \frac{6x^2}{2}$

$\rightarrow x^3 = 3x^2$, divide both side of the equation by $x^2 \rightarrow \frac{x^3}{x^2} = \frac{3x^2}{x^2} \rightarrow x = 3$

# Practice Tests 2:

# Section 1

**1) Answer: C.**

$xp + 4yq = 15 \rightarrow xp = 15 - 4yq$   (1)

$xp + 3yq = 9$   (2)

(1) in (2) $\rightarrow 15 - 4yq + 3yq = 9 \rightarrow 15 - yq = 9 \rightarrow yq = 15 - 9 = 6$

**2) Answer: A.**

$3x^4 + n = a(x^2 + 4)(x^2 - 4) = ax^4 - 16a \rightarrow a = 3$

And $n = -16a = -16 \times 3 = -48$

**3) Answer: A.**

$5x - 4 = 9.5 \rightarrow 5x = 9.5 + 4 = 13.5 \rightarrow x = \dfrac{13.5}{5} = 2.7$

Then; $2x + 3 = 2(2.7) + 3 = 5.4 + 3 = 8.4$

**4) Answer: B.**

$f(x) = x^2 + 3x - 7$

$f(4t^2) = (4t^2)^2 + 3(4t^2) - 7 = 16t^4 + 12t^2 - 7$

**5) Answer: D.**

Let $x$ be all expenses, then $\dfrac{25}{100}x = \$550 \rightarrow x = \dfrac{100 \times \$550}{25} = \$2,200$

He spent for his rent: $\dfrac{22}{100} \times \$2,200 = \$484$

**6) Answer: C.**

$\dfrac{7}{8} = 0.875 \rightarrow C = 5$

$\dfrac{1}{20} = 0.02 \rightarrow D = 2 \rightarrow C \times D = 5 \times 2 = 10$

**7) Answer: C.**

Let P be circumference of circle A, then; $2\pi r_A = 16\pi \rightarrow r_A = 8$

$r_A = 4r_B \rightarrow r_B = \dfrac{8}{4} = 2 \rightarrow$ Area of circle B is $\pi r_B^2 = 4\pi$

**8) Answer: C.**

$y$ is the intersection of the three circles. Therefore, it must be even (from circle A),

negative (from circle B), and multiple of 6 (from circle C).

From the options, only $-18$ is odd, negative and multiple of 6.

**9) Answer: D.**

let $x$ be total number of cards in the box, then number of red cards is: $x - 144$

The probability of choosing a red card is one third. Then: probability$= \frac{1}{3} = \frac{x-144}{x}$

Use cross multiplication to solve for $x$.

$x \times 1 = 3(x - 144) \rightarrow x = 3x - 432 \rightarrow 3x - x = 432 \rightarrow x = 216$

**10) Answer: C.**

Plug in the values of x in each equation and check.

I.  $(-2)^2 - 3(-2) + 5 = 4 + 6 + 5 = 15 \neq 0$

   $(1)^2 - 3(1) + 5 = 1 - 3 + 5 = 3 \neq 0$

II.  $3(-2)^2 + 3(-2) = 12 - 6 = 6 \rightarrow 6 = 6$

   $3(1)^2 + 3(1) = 3 + 3 = 6 \rightarrow 6 = 6$

III.  $2(-2)^2 + 2(-2) - 4 = 8 - 4 - 4 = 0 = 0$

   $2(1)^2 + 2(1) - 4 = 2 + 2 - 4 = 0 = 0$

Equations II and III are correct.

**11) Answer: B.**

Number of biology book: 50

Total number of books; $50 + 68 + 82 = 200$

the ratio of the number of biology books to the total number of books is: $\frac{50}{200} = \frac{1}{4}$

**12) Answer: D.**

A. $f(x) = x^2 + 2$   ; if $x = 1 \rightarrow f(1) = (1)^2 + 2 = 1 + 2 = 3 \neq 4$

B. $f(x) = x^2 - 2$; if $x = 1 \rightarrow f(1) = (1)^2 - 2 = 1 - 2 = -1 \neq 4$

C. $f(x) = \sqrt{x + 3}$   if $x = 1 \rightarrow f(1) = \sqrt{1 + 3} = \sqrt{4} = 2 \neq 4$

D. $f(x) = \sqrt{x} + 3$   if $x = 1 \rightarrow f(1) = \sqrt{1} + 3 = 4 = 4$

Choice D is correct.

**13) Answer: B.**

$\alpha = 180° - 128° = 52°$

$\beta = 180° - 115° = 65°$

$x + \alpha + \beta = 180° \rightarrow$

$x = 180° - 52° - 65° = 63°$

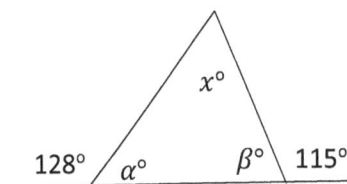

**14) Answer: 190.**

$\frac{16}{100}x = 48 \rightarrow x = \frac{48 \times 100}{16} = 300$

$\frac{1}{5}y = 22 \rightarrow y = 5 \times 22 = 110 \rightarrow x - y = 300 - 110 = 190$

**15) Answer: 36**

Let $b$ be the amount of time Alec can do the job, then,

$\frac{1}{a} + \frac{1}{b} = \frac{1}{30} \rightarrow \frac{1}{180} + \frac{1}{b} = \frac{1}{30} \rightarrow \frac{1}{b} = \frac{1}{30} - \frac{1}{180} = \frac{5}{180} = \frac{1}{36}$

Then: $b = 36$ minutes

**16) Answer: 3.5 or $3\frac{1}{2}$.**

One-degree equals $\frac{\pi}{180}$.

The angle α in radians is equal to the angle α in degrees times π constant divided by 180

degrees. Then: $1\ degree = \frac{\pi}{180}$

$\rightarrow 630\ degrees = \frac{630\pi}{180} = 3.5\pi$

$3.5\pi = x\pi \rightarrow x = 3.5$

**17) Answer: 72.**

In the equilateral triangle if $x$ is length of one side of triangle, then the perimeter of the

triangle is $3x$. Then $3x = 36 \rightarrow x = 12$ and radius of the circle is: $x = 12$

Then, the perimeter of the circle is: $2\pi r = 2\pi(12) = 24\pi$

$\pi = 3 \rightarrow 24\pi = 24 \times 3 = 72$

# Practice Tests 2:

## Section 2

**1) Answer: C.**

$(4^a)^b = 256 \rightarrow 4^{ab} = 256$

$256 = 4^4 \rightarrow 4^{ab} = 4^4 \rightarrow ab = 4$

**2) Answer: C.**

$\sqrt{x} = 5 \rightarrow x = 25$ then; $\sqrt{x} - 9 = \sqrt{25} - 9 = 5 - 9 = -4$

and $\sqrt{x - 9} = \sqrt{25 - 9} = \sqrt{16} = 4$

Then: $(\sqrt{x - 9}) + (\sqrt{x} - 9) = 4 + (-4) = 0$

**3) Answer: C.**

All integers from 15 to 21 are: 15, 16, 17, 18, 19, 20, 21.

The mean of these integers is: $\frac{15 + 16 + 17 + 18 + 19 + 20 + 21}{7} = \frac{126}{7} = 18$

**4) Answer: C.**

$-3a + 5a + 7a = 63 \rightarrow 9a = 63 \rightarrow a = \frac{63}{9} = 7$

Then; $\frac{7a - 1}{6} = \frac{7(7) - 1}{6} = \frac{49 - 1}{6} = 8$

**5) Answer: B.**

$|-16 - 7| - |-12 + 3| = |-23| - |-9| = 23 - 9 = 14$

**6) Answer: B.**

let $x$ be the number of gallons of water the container holds when it is full. Then; $\frac{8}{35}x = 1.6 \rightarrow x = \frac{35 \times 1.6}{8} = 7$

**7) Answer: D.**

Based on the table provided: $g(-2) = g(x = -2) = 3$

$g(4) = g(x = 4) = -4$

$4g(-2) - 2g(4) = 4(3) - 2(-4) = 12 + 8 = 20$

**8) Answer: D.**

Choose a random number for $a$ and check the options. Let $a$ be equal to 21 which is divisible by 7, then:

A. $a - 4 = 21 - 4 = 17$ is not divisible by 5.

B. $a + 4 = 21 + 4 = 25$ is divisible by 5.

  but if $a = 7 \rightarrow a + 4 = 11$ is not divisible by 5.

C. $4a = 4 \times 21 = 84$ is not divisible by 5.

D. $5a - 5 = (5 \times 21) - 5 = 100$ is divisible by 5.

**9) Answer: B.**

$(x - 3)^2 = 9 \rightarrow$ Find the third root of both sides. Then: $x - 3 = 3 \rightarrow x = 6$

$\rightarrow (x - 2)(x - 5) = (6 - 2)(6 - 5) = (4)(1) = 4$

**10) Answer: C.**

The quadrilateral is a trapezoid. Use the formula of the area of trapezoids:

$$Area = \frac{1}{2}h(b_1 + b_2)$$

You can find the height of the trapezoid by finding the difference of the values of $y$ for the points A and D. (or points B and C). $h = 9 - 3 = 6$

AB$= \sqrt{(x_1 - x_2)^2 + (y_1 - y_2)^2} = \sqrt{(6 - 4)^2 + (9 - 9)^2} = \sqrt{4 + 0} = 2$

CD$= \sqrt{(x_1 - x_2)^2 + (y_1 - y_2)^2} = \sqrt{(7 - 1)^2 + (3 - 3)^2} = \sqrt{36 + 0} = 6$

Area of the trapezoid is: $\frac{1}{2}h(b_1 + b_2) = \frac{1}{2}(6)(2 + 6) = 24$

**11) Answer: D.**

First find the number of pants sold in each month.

January: 122, February: 114, March: 120, April: 63, May: 84, June: 72

Check each option provided.

A. There is a decrease from January to February.

B. February and March, $\left(\frac{120-114}{120}\right) \times 100 = \frac{6}{120} \times 100 = 5\%$

C. There is a decrease from March to April.

D. April and May: there is an increase from April to May.

$$\left(\frac{84-63}{84}\right) \times 100 = \frac{21}{84} \times 100 = 25\%$$

**12) Answer: C.**

First, order the number of shirts sold each month: $130, 140, 144, 150, 160, 170$

mean is: $\dfrac{130+140+144+150+160+170}{6} = \dfrac{894}{6} = 149$

Put the number of shoes sold per month in order:

$25, 25, 25, 35, 35, 40$ ; median is: $\dfrac{25+35}{2} = 30$

**13) Answer: B.**

The ratio of number of pants to number of shoes in March equals $\dfrac{120}{40}$.

Three-twelfth of this ratio is $\left(\dfrac{3}{12}\right)\left(\dfrac{120}{40}\right)$. Now, let $x$ be the number of shoes needed to

be added in February.

$\dfrac{114}{35+x} = \left(\dfrac{3}{12}\right)\left(\dfrac{120}{40}\right) \rightarrow \dfrac{114}{35+x} = \dfrac{360}{480} = \dfrac{3}{4} \rightarrow 456 = 3(35+x) \rightarrow 456 = 105 + 3x \rightarrow$

$3x = 351 \rightarrow x = 117$

**14) Answer: C.**

The value of $y$ in the $x$-intercept of a line is zero. Then:

$y = 0 \rightarrow 5x - 3(0) = 8 \rightarrow 5x = 8 \rightarrow x = \dfrac{8}{5}$ then, $x$-intercept of the line is $\dfrac{8}{5}$.

**15) Answer: A.**

Formula of triangle area $= \dfrac{1}{2}$ (base $\times$ height)

Since the angles are $45°, 45°, 90°$, then this is an isosceles triangle, meaning that the

base and height of the triangle are equal.

Triangle area $= \dfrac{1}{2}$ (base $\times$ height) $= \dfrac{1}{2}(10 \times 10) = 50$

**16) Answer: A.**

Let $x$ be the number of years. Therefore, $\$3,000$ per year equals $3,000x$. Starting from

$\$19,000$ annual salary means you should add that amount to $3,000x$.

Income more than that is: $I > 3,000\,x + 19,000$

**17) Answer: D.**

The amount of money for $x$ bookshelf is: $200x$

Then, the total cost of all bookshelves is equal to: $200x + 500$

The total cost, in dollar, per bookshelf is: $\dfrac{Total\ cost}{number\ of\ items} = \dfrac{200x + 500}{x}$

**18) Answer: A.**

Number of Mathematics book: $0.15 \times 500 = 75$

Number of Chemistry book: $0.35 \times 500 = 175$

Product of number of Mathematics and number of Chemistry books: $75 \times 175 = 13,125$

**19) Answer: D.**

According to the chart, 50% of the books are in the Mathematics and Chemistry sections. Therefore, there are 250 books in these two sections: $0.50 \times 500 = 250$

$\gamma + \alpha = 250$, and $\gamma = \dfrac{3}{7}\alpha$ (Replace $\gamma$ by $\dfrac{3}{7}\alpha$ in the first equation)

$\gamma + \alpha = 250 \to \dfrac{3}{7}\alpha + \alpha = 250 \to \dfrac{10}{7}\alpha = 250 \to$ multiply both sides by $\dfrac{7}{10}$ :

$\left(\dfrac{7}{10}\right)\dfrac{10}{7}\alpha = 250 \times \left(\dfrac{7}{10}\right) \to \alpha = \dfrac{250 \times 7}{10} = 175$

$\alpha = 175 \to \gamma = \dfrac{3}{7}\alpha \to \gamma = \dfrac{3}{7} \times 175 = 75$

There are 75 books in the Chemistry section.

**20) Answer: D.**

The angle $\gamma$ is: $0.15 \times 360 = 54°$

The angle $\beta$ is: $0.22 \times 360 = 79.2°$

**21) Answer: D.**

The angles on a straight line add up to 180 degrees. Then:

$x + 10 + 3y + 2x + y = 180$

Then, $3x + 4y = 180 - 10 \to 3(30) + 4y = 170 \to 4y = 170 - 90 = 80 \to y = 20$

**22) Answer: A.**

The distance of A to B on the coordinate plane is: $\sqrt{(x_1 - x_2)^2 + (y_1 - y_2)^2} =$

$\sqrt{(4 - 1)^2 + (7 - 3)^2} = \sqrt{3^2 + 4^2} = \sqrt{9 + 16} = \sqrt{25} = 5$

The diameter of the circle is 5 and the radius of the circle is 2.5. Then: the circumference of the circle is: $2\pi r = 2\pi(2.5) = 5\pi$

**23) Answer: D.**

Square root of 16 is $\sqrt{16} = 4 < 7$

Square root of 9 is $\sqrt{9} = 3 < 7$

Square root of 48 is $\sqrt{48} = \sqrt{49 - 1} < \sqrt{49} = 7$

Square root of 64 is $\sqrt{64} = 8 > 7$

Since, $\sqrt{49} < \sqrt{64}$, then the answer is D.

**24) Answer: C.**

The area of the trapezoid is: $Area = \frac{1}{2}h(b_1 + b_2)$

$A = 104 = \frac{1}{2}(x)(10 + 16) \rightarrow 8x = 64 \rightarrow x = 8$

$y = \sqrt{6^2 + 8^2} = \sqrt{36 + 64} = \sqrt{100} = 10$

The perimeter of the trapezoid is: $8 + 10 + 10 + 16 = 44$

**25) Answer: A.**

$|x - 3| \geq 12$; Then: $x - 3 \geq 12 \rightarrow x \geq 12 + 3 \rightarrow x \geq 15$

Or $x - 3 \leq -12 \rightarrow x \leq -12 + 3 \rightarrow x \leq -9$

Then, the solution is: $x \geq 15 \cup x \leq -9$

**26) Answer: B.**

Since, E is the midpoint of AB, then the area of all triangles DAE, DEF, CFE and CBE are equal.

Let $x$ be the area of one of the triangles, Then: $4x = 120 \rightarrow x = 30$

The area of DEC $= 2x = 2(30) = 60$

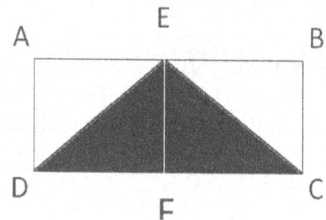

**27) Answer: A.**

$5 < -4x - 3 < 13 \rightarrow$ Add 3 to all sides. $5 + 3 < -4x - 3 + 3 < 13 + 3$

$\rightarrow 8 < -4x < 16 \rightarrow$ Divide all sides by $-4$.

(Remember that when you divide all sides of an inequality by a negative number, the inequality sing will be swapped. $<$ becomes $>$)

$\frac{8}{-4} > \frac{-4x}{-4} > \frac{16}{-6} \rightarrow -2 > x > -4, or\ -4 < x < -2$

**28) Answer: 16.97.**

Based on triangle similarity theorem: $\frac{a}{a+b} = \frac{c}{5} \rightarrow c = \frac{5a}{a+b} = \frac{5\sqrt{2}}{\sqrt{3}+4\sqrt{3}} = 1$

$\rightarrow$ area of shaded region is: $\left(\frac{c+5}{2}\right)(b) = \frac{6}{2} \times 4\sqrt{3} = 12\sqrt{2}$

Round $12\sqrt{2}$ to the nearest hundredths place gives 16.97.

**29) Answer: 100.**

One liter=1,000 cm$^3 \rightarrow$ 9 liters $= 9,000$ cm$^3$

$9,000 = 18 \times 5 \times h \rightarrow h = \frac{9,000}{90} = 100$ cm

**30) Answer: 9.**

Based on corresponding members of each matrix, write two equations:

$\begin{cases} 4x = 3x + 2y - 3 \\ 5x = 8y - 9 \end{cases} \rightarrow \begin{cases} x - 2y = -3 \\ 5x - 8y = -9 \end{cases}$

Multiply first equation by $-5$.

$\begin{cases} -5x + 10y = 15 \\ 5x - 8y = -9 \end{cases} \rightarrow$ add two equations.

$2y = 6 \rightarrow y = 3 \rightarrow x = 3 \rightarrow x \times y = 9$

**31) Answer: $\frac{\sqrt{2}}{2}$.**

$\cos(A) = \frac{adjecent}{hypotenuse} = \frac{\sqrt{2}}{2} \Rightarrow$ We have the following triangle, then:

$b = \sqrt{2^2 - (\sqrt{2})^2} = \sqrt{4-2} = \sqrt{2}$

$\sin(A) = \frac{\sqrt{2}}{2}$

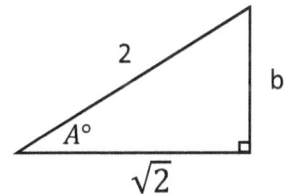

**"End"**

www.ingramcontent.com/pod-product-compliance
Lightning Source LLC
Chambersburg PA
CBHW081325090426

42737CB00017B/3036